S 373.36 AND
ITEM NO: 1961720
19617
D0491498

AQA

Design and Technology

Electronic Products

GCSE

UNIVERSITY OF WALES, NEWPORT
LIBRARY AND INFORMATION SERVICES
CAERLEON

Paul Anderson
Neil Cafferky
Samantha Forsyth
Richard Johnson
Harry Longworth
Keith Mellens

UNIVERSITY OF WALES, NEWPORT
LIBRARY AND INFORMATION SERVICES
CAERLEON

Nelson Thornes

1401720

Text © Paul Anderson and Nelson Thornes 2009
Original illustrations © Nelson Thornes Ltd 2009

The right of Paul Anderson to be identified as author of this work has been asserted by him in accordance with the Copyright, Designs and Patents Act 1988.

All rights reserved. No part of this publication may be reproduced or transmitted in any form or by any means, electronic or mechanical, including photocopy, recording or any information storage and retrieval system, without permission in writing from the publisher or under licence from the Copyright Licensing Agency Limited, of Saffron House, 6-10 Kirby Street, London EC1N 8TS.

Any person who commits any unauthorised act in relation to this publication may be liable to criminal prosecution and civil claims for damages.

Published in 2009 by:
Nelson Thornes Ltd
Delta Place
27 Bath Road
CHELTENHAM
GL53 7TH
United Kingdom

09 10 11 12 13 / 10 9 8 7 6 5 4 3 2 1

A catalogue record for this book is available from the British Library

ISBN 978 1 4085 0417 8

Cover photograph by Jim Wileman DISCLAIMER: Suitable personal protective equipment should always be worn.

Page make-up by Hart McLeod, Cambridge
Artwork by Hart McLeod, Cambridge.
Index compiled by Indexing Specialists (UK) Ltd
Printed in Great Britain by Scotprint

Acknowledgments

The authors and publisher are grateful to the following for permission to reproduce the following copyright material:

Chapter strips: Chapter 1: iStockphoto.com/© Yury Kosourov; Chapter 2: iStockphoto.com/© David Coder; Chapter 3: iStockphoto.com/© Cristina Dumitras; Chapter 4: iStockphoto.com/ © MH; Chapter 5: iStockphoto.com; Chapter 6: iStockphoto.com/© Nancy Nehring; Chapter 7: iStockphoto.com/© Dmitry Nikolaev

Alamy 83; Fotolia.com 9, 14, 15, 20, 64, 73, 78, 79, 80, 86, 97; Gettyimages 8; iStockphoto 13, 58, 75, 94; Jim Wileman 6, 7; Maplin.co.uk 14, 15, 16, 19; Rapidonline.com 14, 15, 16, 17, 18, 21, 68; Science Photo Library 90, 91.

Special appreciation is offered to Balcarras School, Cheltenham and Great Barr School, Birmingham.

Every effort has been made to contact the copyright holders and we apologise if any have been overlooked. Should copyright have been unwittingly infringed in this book, the owners should contact the publishers, who will make corrections at reprint.

The Controlled Assessment tasks in this book are designed to help you prepare for the tasks your teacher will give you. The tasks in this book are not designed to test you formally and you cannot use them as your own Controlled Assessment tasks for AQA. Your teacher will not be able to give you as much help with your tasks for AQA as we have given with the tasks in this book.

Contents

Nelson Thornes and AQA

Nelson Thornes has worked in partnership with AQA to ensure this book and the accompanying online resources offer you the best support for your GCSE course.

All resources have been approved by senior AQA examiners so you can feel assured that they closely match the specification for this subject and provide you with everything you need to prepare successfully for your exams.

These print and online resources together **unlock blended learning**; this means that the links between the activities in the book and the activities online blend together to maximise your understanding of a topic and help you achieve your potential.

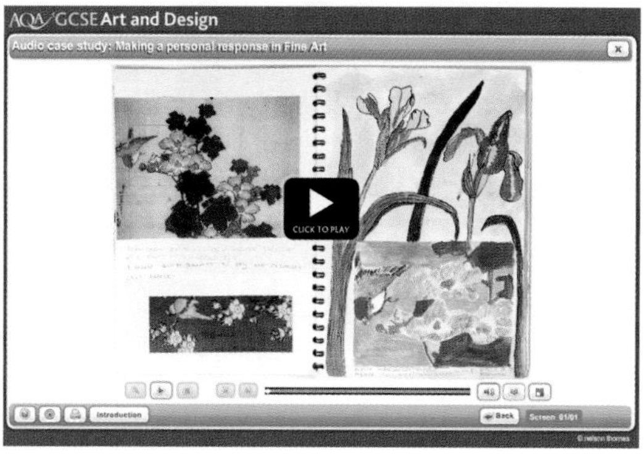

These online resources are available on kerboodle! which can be accessed via the internet at **www.kerboodle.com/live**, anytime, anywhere. If your school or college subscribes to kerboodle! you will be provided with your own personal login details. Once logged in, access your course and locate the required activity.

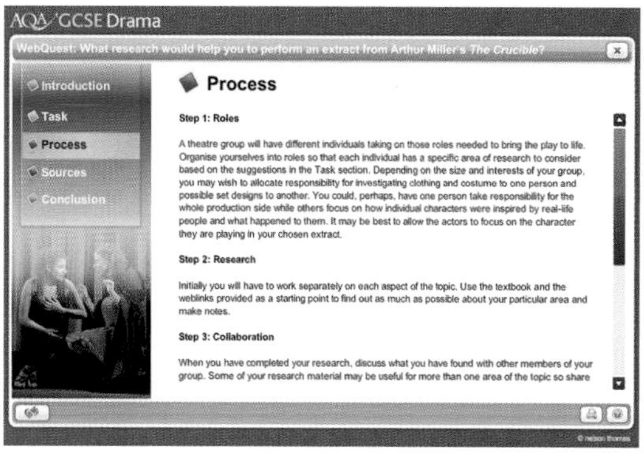

For more information and help on how to use kerboodle! visit **www.kerboodle.com**.

How to use this book

Objectives

Look for the list of **Learning Objectives** based on the requirements of this course so you can ensure you are covering everything you need to know for the exam.

AQA Examiner's tip

Don't forget to read the **AQA Examiner's Tips** throughout the book as well as practise answering **Examination-style Questions**.

Visit **www.nelsonthornes.com/aqagcse** for more information.

AQA examination-style questions are reproduced by permission of the Assessment and Qualifications Alliance.

A *A student modelling a circuit*

Introduction

The book structure

This book is divided into two units which correspond to the units in the GCSE Design and Technology Electronic Products specification.

Unit 1 looks at the following topics which will be tested in the written paper:

- Materials and components
- Design and market influences
- Processes and manufacture

Unit 2 provides guidance on how to be successful with the controlled assessment unit.

This book will fully prepare you for the GCSE Design and Technology Electronic Products course. You will have all the knowledge that you will need to succeed in the written examination and you will be able to test yourself with examination-style questions. You will be carefully led through the demands of the Controlled Assessment task. There are examples of high quality students' work together with detailed commentary and tips from the team of examiners.

Electronic Products

Our world is full of electrical and electronic products. These range from the alarm clocks that wake us in the morning, to the fridges that store our food, the radios and MP3 players we listen to, the computers and mobile phones we use, the control systems in the cars and buses we travel in… From the moment we wake in a morning to the time we return to bed, our whole day will be influenced and affected by these products. They help to shape our lifestyle and also define us as the people we are. These products are usually made in large quantities and have to fill a lot of different, sometimes conflicting, design needs. They must be manufactured to an appropriate quality and presented for sale at a suitable price. To achieve this designers have to make many decisions with regard to how these products will function, their styling, the materials used and how they will be made. They also have to take account of moral, ethical and sustainable issues. Electronic Products is complex and it is hoped that by reading this book you will be fully aware of the issues affecting this subject.

Designing

To be a good designer of electronic products you need to know about electrical and electronic components. You will learn about the functions of a wide range of components and how they can be used together to satisfy a design need.

You need to be able to take account of all of the different design needs to generate ideas for both electronic systems and the enclosures needed to house them. You will learn techniques that will help you become a creative designer and learn about different methods of presenting your ideas to others.

A designer must consider the impact their product is likely to have on others. In this country we consume far more of the world's resources than we should and many of the products we buy exploit people in other countries. By reading this book, it is hoped that you will not only become a better designer but a more informed consumer.

Making

If you are to develop your design into a working product, you will also need to know about the materials used to make the enclosures. You will learn about the advantages and disadvantages of using a variety of materials. You need to know about different methods of manufacture. You will learn how to cut, shape, form, mould, condition, assemble and finish a range of materials. You will also learn how to work safely and how to use industrial methods of manufacture to improve the accuracy and consistency of the products you design.

....and finally

Electronic Products is an exciting and very rewarding course. It will involve you in a great deal of decision making and hard work. You will need to plan ahead and become very organised. In the end, you should finish up with a wide range of knowledge, skills and understanding that will be useful to you over the coming years, and hopefully, you will have designed and made at least one product that you can be really proud of.

B *Students making circuits*

UNIT ONE Written paper

Systems and components

What will you study in this section?

After completing Systems and components (chapters 1–4) you should have a good understanding of:

- how electronic products work
- inputs
- processes
- outputs.

Key terms

Electrical: electrical components are simple conductors and perform a function when electricity flows through them.

Electronic: electronic components are devices that include semiconductor materials in an electrical circuit.

Component: an individual part.

Circuit: an assembly of electrical or electronic components that exists to perform a function.

Function: what the product is intended to do.

How electronic products work

There are a huge number of **electrical** and **electronic** products in everyday use, ranging from the automatic doors in shops to kettles and computers to MP3 players. Many of these products have been invented within the last 50 years. Since their initial development, they have been continually improved, leading to better performance, reduced size or lower cost.

Consider, for example, mobile phones. The first commercial mobile phone service was launched in Japan in 1978. The first mobile phones were large and bulky, Photo **A**. The handset was often bigger and heavier than a house brick. The size and weight was necessary due to the electrical **components** used to make it and the battery. The only facility they allowed was to make and receive calls.

Over time, improvements in electronic components, battery technology and manufacturing have allowed the phone to shrink in size. In addition, the smaller components have allowed more components to be included in the circuits, giving additional capabilities. Mobile phones can now send texts, take pictures and play music, giving more features to the user. The limiting factor on size is now no longer the components and the **circuit** – it is the need to be big enough so that it can be operated by hand.

A *How current mobile phones compare to early models*

Each of these new and improved products needed circuits to be designed. The aim of this section is to support the development of electrical and electronics knowledge that will enable you to understand and design circuits. To do this, we first need to understand how circuits are designed.

How modern electronic products are created

Modern electronic circuits can be very complicated. They may contain hundreds of components. If these circuits had to be completely designed by hand, it would take a lot of skill and a long time. The designer would have to select and position every component in the circuit and carry out lots of calculations to make sure that every part would work together. Fortunately, there is an alternative, called systems analysis. In this, the circuit designer first decides what **functions** the circuit must carry out. He or she then selects standard circuit blocks which would be able to carry out these functions. These blocks are existing, tested sub-circuits. By putting together the blocks needed for a circuit, the designer can very quickly create a draft design.

The next step is to modify these standard blocks slightly, to make sure that the circuit design can exactly meet the individual needs of the application. This may involve substituting some components with alternatives or changing the values of some components. This, in turn, may need calculations to be carried out to check that the new components will achieve the required results. Finally, the circuit designs may be tested and modelled, and a prototype circuit built.

The approach to circuit design in GCSE Electronic Products is very similar to this. You should be able to design simple electrical circuits from start to finish. However, for complicated electronic circuits it is acceptable to start with existing circuit designs, or circuit designs generated using computer software, and then to customise them so that they are suitable to meet the needs of the product that you are designing. To do this, you need to understand what each of the components used in the different systems blocks does. In this chapter, we will investigate each of the different components, identifying what they are used for and how they can be used in combination with other components. We will also investigate the function of some of the simple systems blocks that might be used in a typical GCSE project.

Remember

Electricity is a form of energy. It involves the flow of electrons around a circuit. The rate of flow of this electric charge is called the current, measured in amperes. The driving force for this flow of charge is called the potential difference or voltage.

B *MP3 players are a modern electronic product*

Activity

Make a list of all the items of electrical equipment that you have used, or benefited from, in the last week.

1 Systems and components

1.1 Systems diagrams 1

Objectives

- Explain what a system is.
- List the parts of a systems diagram.
- Explain how a systems diagram is used.

Key terms

System: a collection of parts that interact with their environment and perform a function.

Systems diagram: a schematic representation of a system.

Systems diagrams

A **system** is a collection of parts that exists to perform a function. Any product that is an assembly of parts and interacts with its environment in some way is a system.

A **systems diagram** is a representation of how a system will work. It breaks down the functions to be carried out into simple categories and lists the parts needed to carry out that category. This means that if someone can create a systems diagram, they can produce an overview of the functional blocks needed to design any product that is a system.

What are the parts of a system?

Diagram **A** shows the parts of a simple system. The power supply (for example the battery or mains electricity) is never shown as part of the system.

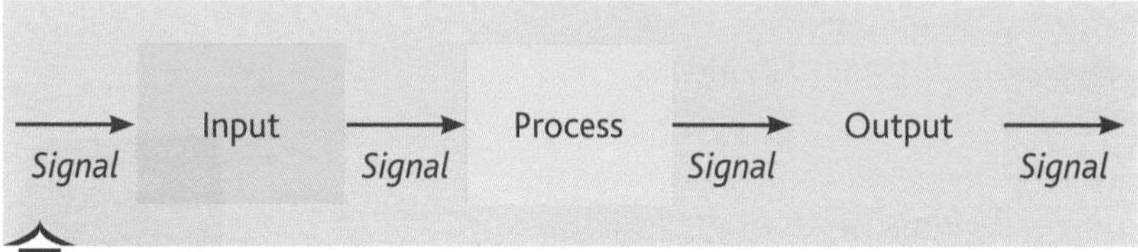

A *Systems diagram*

The signal

A very common mistake when creating systems diagrams is to overlook the signal. Every box must be connected by a signal. The signal shows the route into, through and out of the system.

The signal must be measurable. However, the type of signal can be different between boxes. Common signals include movement, electricity, light and sound.

The process boxes

Every process box in the system must have a function. This means that it must change the signal in some way. For example, it could change the size of the signal, perhaps by increasing or reducing the electrical current. Alternatively, it could change the type of signal, perhaps by converting electricity into light or air pressure into movement.

Each of the process boxes is normally a physical item or, more often, an assembly of components. A simple system may have just one process box of each type, for example Diagram **B**. Complicated systems may have several process boxes, for example Diagram **C**.

Remember

Examples of systems

- Cars
- Engines
- Vending machines
- Mobile phones
- MP3 players
- Computers

Input

The input is the thing that starts the system. This is normally some form of sensor. For example, the inputs in a simple house alarm may include an on-off switch, a keypad for entering the alarm code, a contact switch on a door and motion sensors in the rooms.

Process

The process is the brain of the system. It says what should happen when the system is activated. In the case of the alarm, this may include a latching function so that when a sensor is activated the alarm stays on and a timer function so that the siren will stop after 5 minutes, to avoid upsetting the neighbours. These could both be achieved with a single electronic microcontroller (PIC or picaxe) circuit. There are certain types of process box that are commonly used in electronic products. These include:

- comparators – these compare a signal to a reference value
- latches – a switch that stays on once activated
- logic gates – for example AND, OR and NOT gates. These will be explained later
- counters
- timers
- pulse generators
- drivers – a driver is a block that increases the output power so that it is able to operate an output.

Output

The output is how the system responds to being activated. For example, for the alarm, it may be a siren and a flashing light, to convert the electrical signal from the process box to sound and light signals respectively.

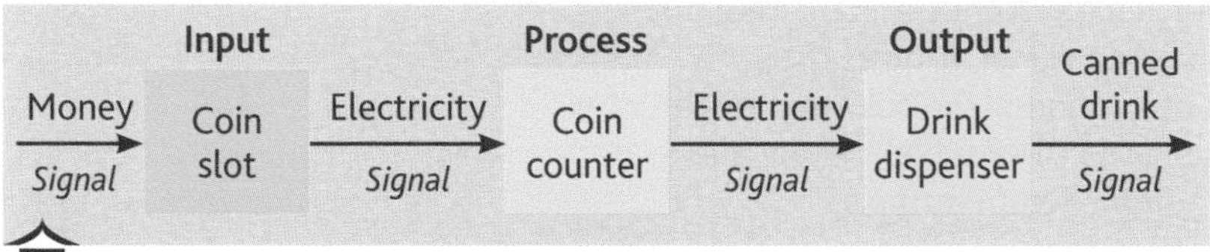

B *Example of a systems diagram for a drinks machine*

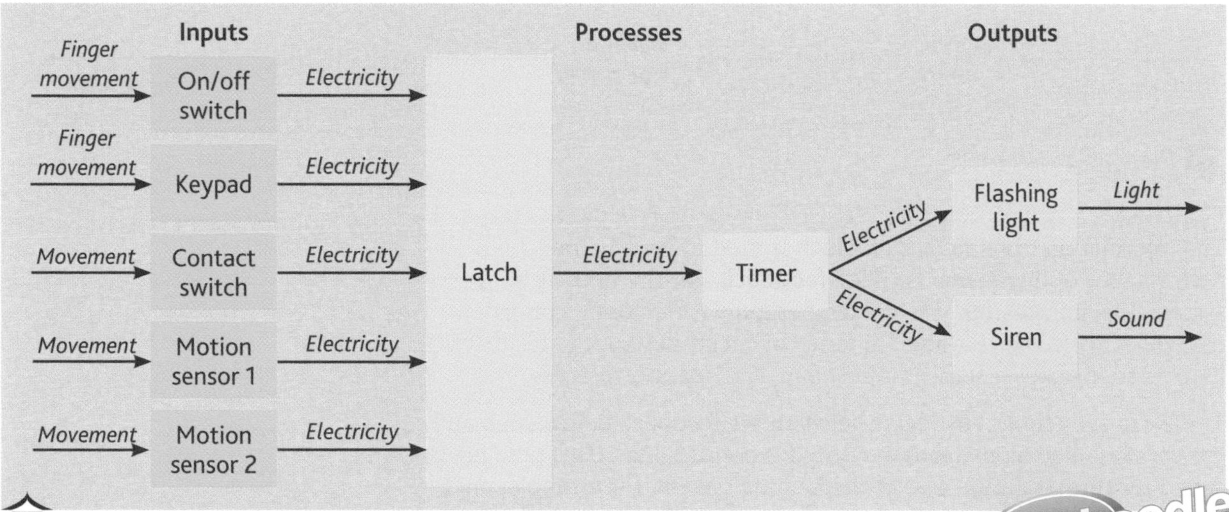

C *Systems diagram for a house alarm*

Activities

1. Produce a systems diagram for a security light that sits outside a garage. It should be activated when an intruder comes near and should stay on for 15 minutes. Clearly label the signals, inputs, processes and outputs.
2. Produce a systems diagram for a laptop computer.

AQA Examiner's tip

Use a systems diagram to work out the parts that need to be included in your design.

Summary

A system is a collection of parts that exists to perform a function.

The parts of a system are signals, inputs, processes and outputs. A complex system may have several process boxes of each type.

A systems diagram is a useful tool to help identify the functional parts that need to be included in a product.

1.2 Systems diagrams 2

Objectives

Recognise open and closed loop feedback systems.

Key terms

Open loop: a system that is set to achieve a required value.

Closed loop: a system that can alter its output based on feedback.

Feedback: information from sensors used to modify the output of a system.

Hysteresis: the time lag between a correction being made to a system and the output of the system returning to the target value.

Types of systems diagram

Systems can be broadly classified into two types: **open loop** or **closed loop**.

Open loop

An open loop system is one that is set up with the intention of achieving a certain outcome or result, but that has no way of checking that the result has been achieved.

An example of an open loop system is an old-fashioned central heating system. This might consist of an on-off switch, a boiler (including a pump) and some radiators, Diagram **A**. The aim of the system would be to heat one or more rooms. When you were cold, you would activate the switch and the heating would turn on. The heating would stay on until it was turned off, whether the room remained cold or became unbearably hot.

Closed loop

In a closed loop system, the system is able to check whether it is meeting the required outcome or result and correct itself if necessary, Diagram **B**.

This is normally achieved by adding an additional sensor that monitors the output condition. For example, considering the central heating system from above, you could add a device called a thermostat, Photo **C**. This is set at the required temperature and continually checks the temperature in the room. If the room is too cold, it turns the radiator on, allowing it to be heated. If the room is too warm, it turns the radiator off. Compared to the extremes of too hot or too cold that might be experienced with the open loop system, the closed loop system allows the room to be kept at a constant, comfortable temperature.

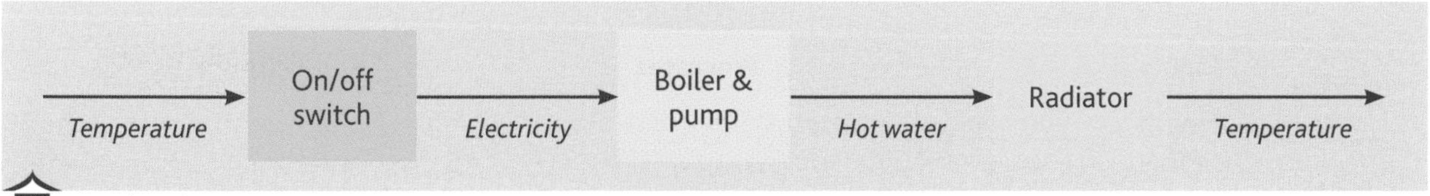

A *Open loop heating system*

The information from an output which is used to modify the performance of the system is called **feedback**. For the heating system, this information was the temperature in the room, compared electronically to the set value. In terms of circuit design, the feedback loop often includes a sensor (input) and a comparator process.

There can sometimes be a delay between the feedback being given and the correction being made to the system's performance. This time lag is called **hysteresis**. In the case of the heating system, for example, once the thermostat turns the heating on, the boiler will heat and pump the water and the air in the room will slowly rise to temperature. This hysteresis is one of the reasons that a manufacturer may set a device to a precise value, but then quote a range of output performance. For example, the heating manufacturer could set the temperature to be maintained at 23°C but state that the temperature could vary by 3°C either way. This means that, allowing for hysteresis, the room in this example may be between 20°C and 26°C.

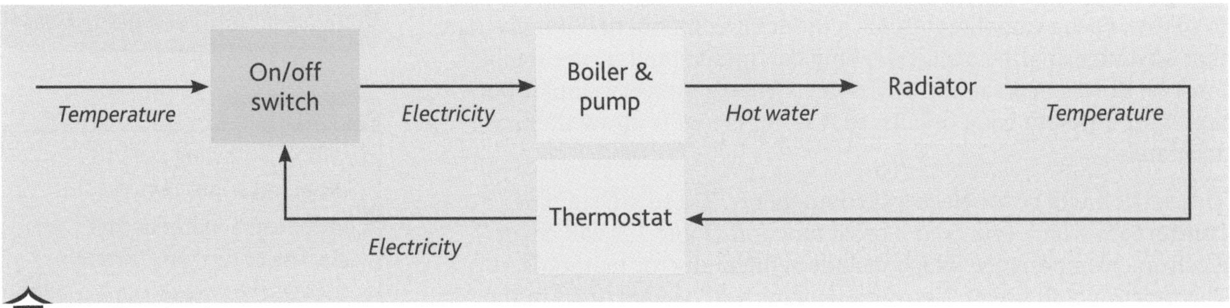

B *Closed loop heating system*

C *Electronic thermostat*

Activity

Produce a systems diagram for a filter coffee maker. Once turned on, it should heat the water in the machine for 5 minutes, then pump it through the coffee into a jug on a hot plate. The hot plate should be held at 90°C, to make sure that the coffee is always warm when it is poured.

AQA Examiner's tip

On your systems diagram, you should show any feedback that is taking place and explain how this will be achieved.

Summary

An open loop system is set up to achieve an outcome, but does not check that it has achieved what is required.

A closed loop system uses feedback to modify its performance, so that it achieves the required outcome. There may be a time lag, called hysteresis, between the system modifying itself and the required outcome being achieved.

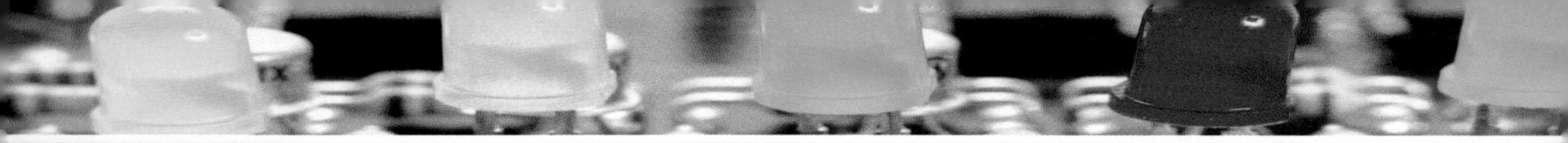

1.3 Circuit symbols 1

Circuit symbols

Electrical circuits are drawn using **symbols**, joined together by lines to represent the connections. Symbols make the diagram much easier to follow. Each symbol represents a different component. To make sure that anyone using the circuit drawing can understand it, the symbols must be drawn in a certain way. Some symbols are very similar but represent different components, so it is important to draw them accurately.

This section will cover electrical components. These are simple **conductors** which carry out a basic function. Topic 1.5 will cover electronic components, which include semiconductor materials. The function and use of these components will be explained in the following sections.

A *Power supplies*

Cell		
Battery	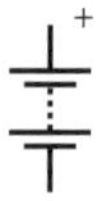	
Voltage rail	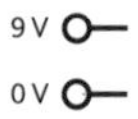	

B *Resistors*

Resistor		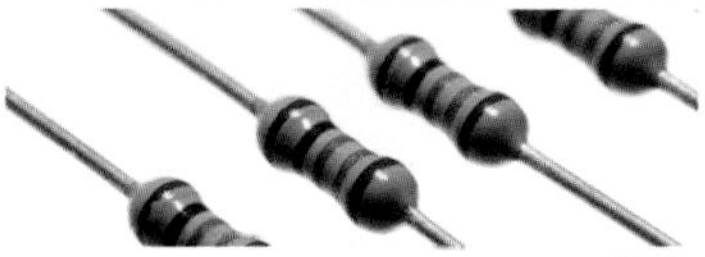
Variable resistor		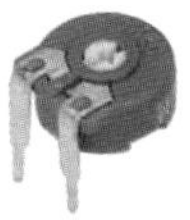
Potentiometer	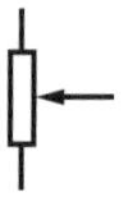	
Thermistor		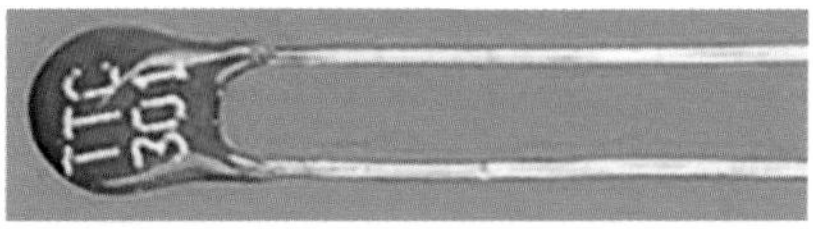
Light-dependent resistor		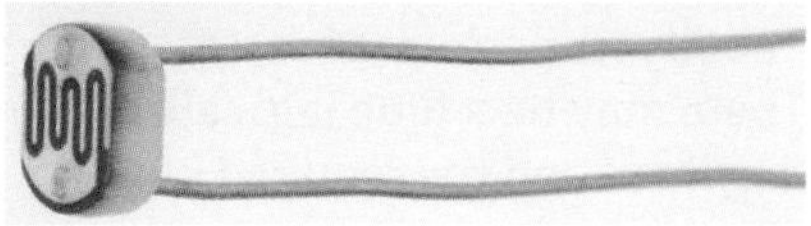

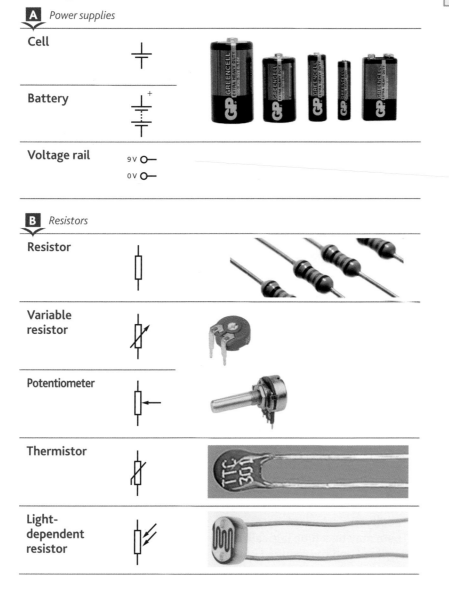

Objectives

Recognise the symbols for electrical components used in circuit diagrams.

Key terms

Symbol: a drawing used to represent a component.

Conductor: a material that electricity can pass through, such as the wire or track used to connect components.

C *Capacitors*

Capacitor	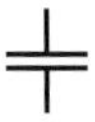	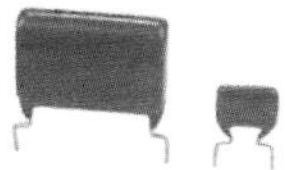
Electrolytic capacitor		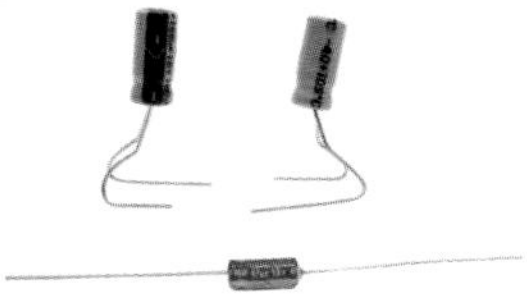

D *Switches*

Push-to-make switch		
Push-to-break switch		
Single-pole single-throw switch		
Single-pole double-throw switch		

E *Light-emitting diodes*

Light-emitting diode		
Flashing light-emitting diode		
Bi-colour light-emitting diode	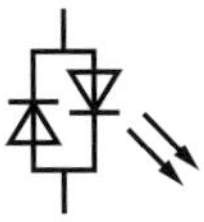	
Tri-colour light-emitting diode		

F *Other output devices*

Buzzer		
Loudspeaker		
Bell		
Microphone	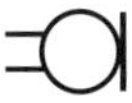	
Lamp		
Motor		

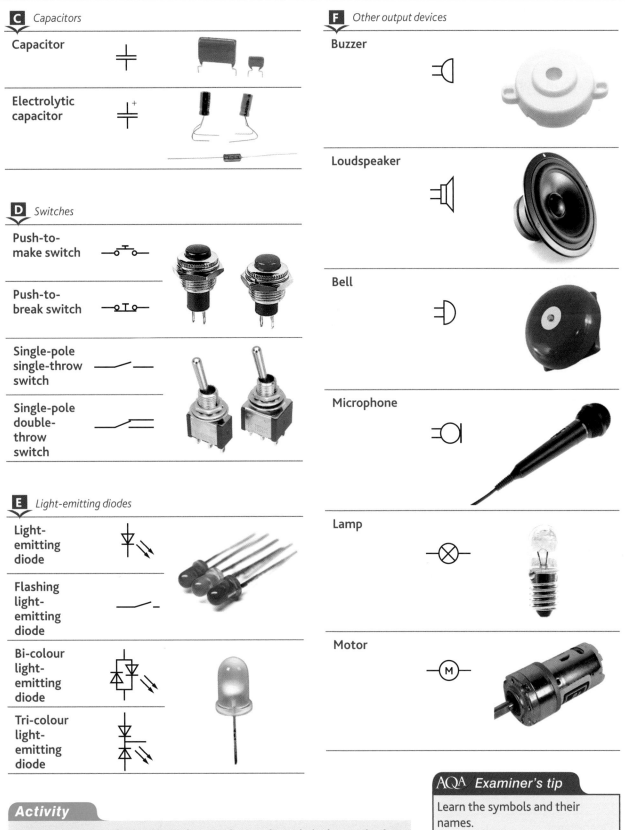

Activity

Using the internet, find a circuit diagram for a radio. Label what each of the symbols used represents.

AQA Examiner's tip

Learn the symbols and their names.

kerboodle!

1.4 Circuit symbols 2

Circuit symbols

This section will cover electronic components which include semiconductor materials. Topic 1.3 covered electrical components. These are simple conductors which carry out a basic function.

The function and use of these components will be explained in the following sections.

Objectives

Recognise the symbols for electronic components used in circuit diagrams.

Draw circuit diagrams.

Key terms

IC: integrated circuit.

A *Semi-conductors*

Photo transistor

Diode

NPN transistor

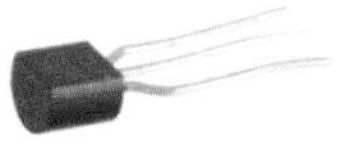

Field effect transistor

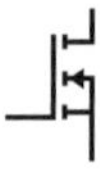

Opto-isolator

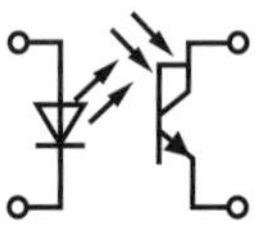

Piezocrystal oscillator

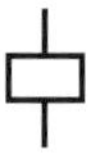

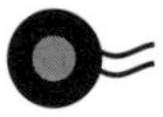

Thyristor

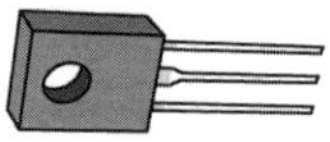

There are many types of **IC**, but these are the ones you need to be able to recognise.

B *Integrated circuits (ICs)*

555 timer IC

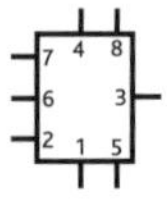

Operational amplifier (op amp)

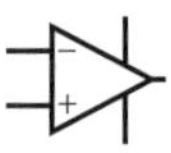

Voltage regulator

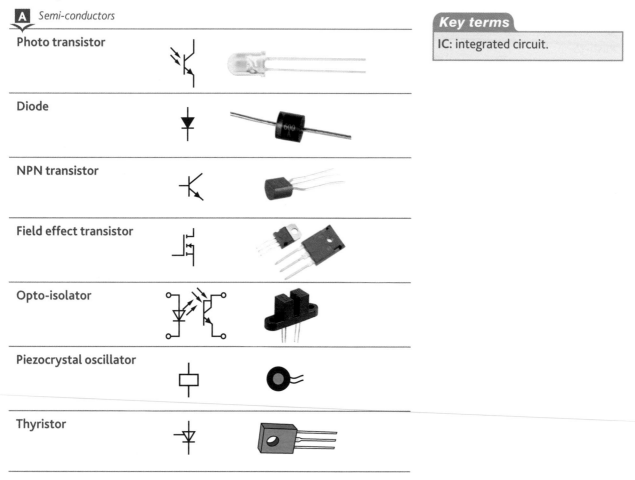

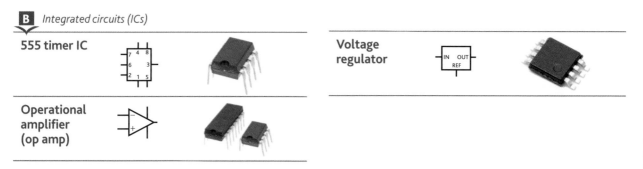

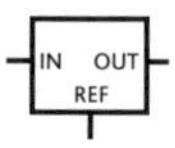

The logic gates used in electronic circuits are normally ICs. These are the ones you need to be able to recognise.

Circuit diagrams

A circuit diagram is a schematic illustration of a circuit. This means that the diagram shows how the components are linked together, but may bear little resemblance to what the parts look like in the real world.

When you draw a circuit, always draw straight lines. Lines should be either vertical or horizontal, not sloping.

Where lines cross, a dot is used to indicate that they are connected. If lines cross without a dot it means that they are not connected, Diagram **D**.

There are a number of software packages to help you draw circuit diagrams. Some packages can also simulate the circuit so that you can try it out and see it working.

C *Logic gates*

AND gate	
OR gate	
NOT gate	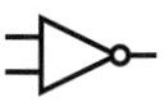

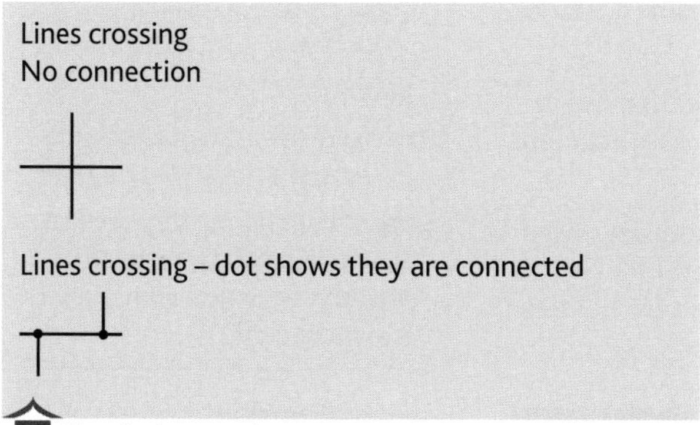

D *Electrical connections*

Example circuit

Diagram **E** shows a circuit that turns a lamp on or off when the light level in a room changes.

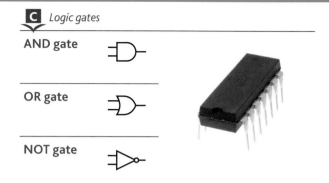

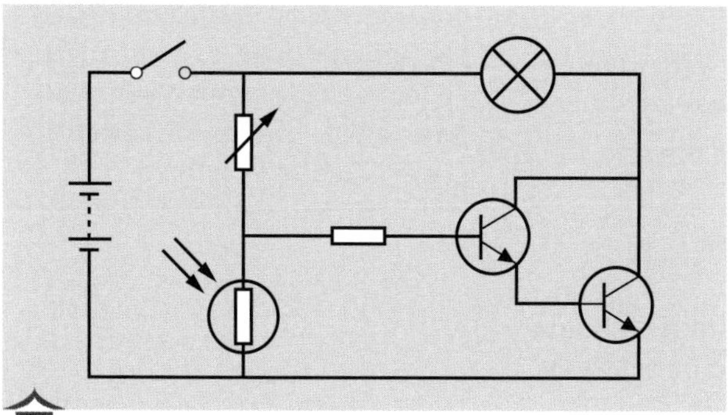

E *Example of a circuit diagram*

Activities

1. On Diagram **E**, label the individual components used.
2. Using the internet, find a circuit diagram for an alarm or a similar circuit including an integrated circuit. Label what each of the symbols used represents.

AQA Examiner's tip

- Learn the symbols and their names.
- Always use a ruler for connection lines.
- Make sure connections are clearly shown.

kerboodle!

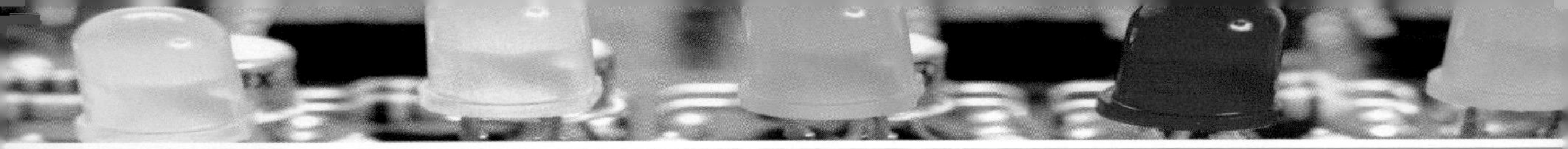

1.5 Batteries and diodes

Batteries

The power supply is not shown on systems diagrams, but power is assumed to be available. The most common source of electricity used for most GCSE Electronic Products projects are **batteries**. These are much safer than mains supplies and also allow the circuits manufactured to be portable.

Characteristics of batteries

Batteries store electricity in the form of chemical energy. The chemicals in the battery react together to create the electricity. Once all of the chemicals have been used, the battery stops producing electricity.

Batteries can be classified according to the chemicals they contain. Zinc carbon batteries are typically the cheapest. Alkaline batteries offer longer life, with more amp hours and higher capacity. Rechargeable batteries typically contain nickel-cadmium or nickel-metal hydride. Rechargeable batteries cost more than zinc carbon or alkaline batteries, but they can be recharged many times.

Silver oxide button cells and lithium disc cells are designed for use in watches, calculators and other small electronic products. These are not typically used in GCSE projects.

Batteries come in a wide variety of sizes, Photo **A**. Common sizes used in GCSE projects include AAA, AA, C and D (all 1.5 V) and PP3 (9 V). Batteries of the same size can be combined in series in battery holders to produce supplies of 3 V, 4.5 V, 6 V, 9 V and 12 V.

A *D, C, AA, AAA and PP3 batteries*

Choosing a suitable power supply

When choosing a battery, the following need to be considered:

- Performance – what voltage and current are required? What battery life (amp hours) is needed?
- Physical characteristics – what limitations are there on the size, shape or weight?
- Cost – is it more cost-effective in the longer term to use rechargeable or non-rechargeable batteries?
- Disposal – when they have been drained of charge, batteries should not be thrown away with domestic waste. They should be sent to a dedicated recycling centre.

One environmentally friendly alternative to batteries is photovoltaic cells. These generate current from sunlight. They are not yet a practical alternative to batteries for powering circuits, but in the near future they will almost certainly be a sustainable alternative.

Objectives

- Describe the characteristics of a range of batteries.
- Explain the considerations for selecting an appropriate power supply.
- Describe how a diode can be used to prevent damage to a circuit.

Key terms

Battery: a component or unit which stores electrical energy chemically.

Diode: a component that allows current to flow in one direction only.

Polarised: having a positive side or leg and a negative side or leg.

Back emf: a momentary reverse flow of electricity when an electromechanical component is switched off.

Remember

Product design and batteries

The design of the finished product in your GCSE project needs to consider the batteries used. For example, the batteries should be located safely and securely. Access may be needed to replace or recharge them, as appropriate. It may also be necessary to allow the voltage level of the battery to be checked or monitored.

Diodes

Many electronic components can be damaged if the battery is connected to them the wrong way round. A **diode**, Photo **B**, is a component which allows current to flow through it in one direction only, indicated by the direction of the arrow on the circuit symbol. It must be connected the correct way round in a circuit for it to conduct electricity. Diodes are used to prevent damage to components in a wide range of different systems blocks.

How a diode works

A diode is a **polarised** component. It has a positive end, called the anode, and a negative end, called the cathode. Often, on the actual component a silver or coloured band identifies the cathode end. If the anode of the diode is connected to a higher voltage than the cathode, current will flow from anode to cathode. This is called forward bias. If the diode is reversed, so that the cathode is at a more positive voltage than the anode, the diode will not conduct electricity. This is called reverse bias.

Using diodes to prevent damage to circuits

Diagram **C** shows how a lamp in a circuit only lights when the diode is forward biased. A diode can be similarly placed in other circuits to prevent the risk of damage from a battery being connected the wrong way round. Circuits can also be damaged by **back emf.** This is a momentary reverse in the flow of electricity that can occur when components such as relays, solenoids or motors are switched off. Diagram **D** shows a diode connected in reverse bias to protect the transistor from back emf when the relay is switched off. A diode used in this way is called a clamping diode.

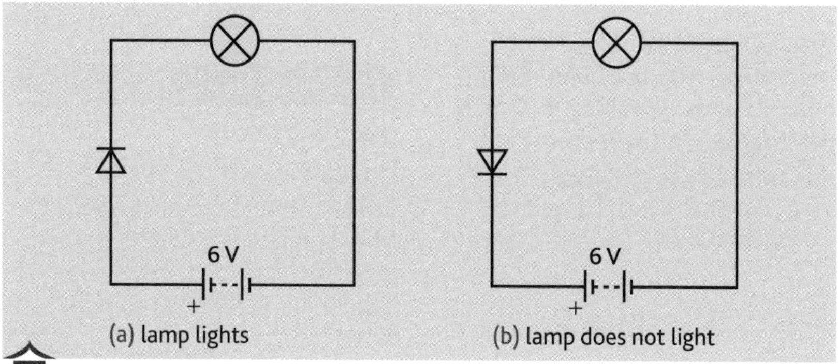

C *Effect of diode direction on a circuit*

Summary

Batteries are available in a range of different sizes and types, to suit the needs of the application.

Factors to consider when selecting a battery for a new circuit include the performance, physical characteristics, cost and disposal.

A diode only allows current to pass in one direction.

A diode can be used to prevent damage in circuits from the battery being attached incorrectly or due to back emf from electromechanical components.

Remember

Battery life

Battery life is measured in milliamp hours. A 9 V NiCad battery has a life of 110 mAh. This means that if the battery is driving a circuit which needs 110 mA it will last one hour. If it were driving an LED which needs 25 mA to work it would last approximately 4 hours.

Activity

Create a table of the different battery types and sizes, listing their performance, physical characteristics and cost.

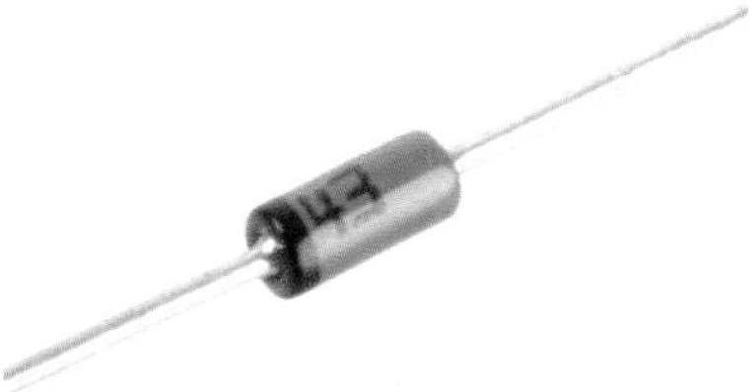

B *A diode*

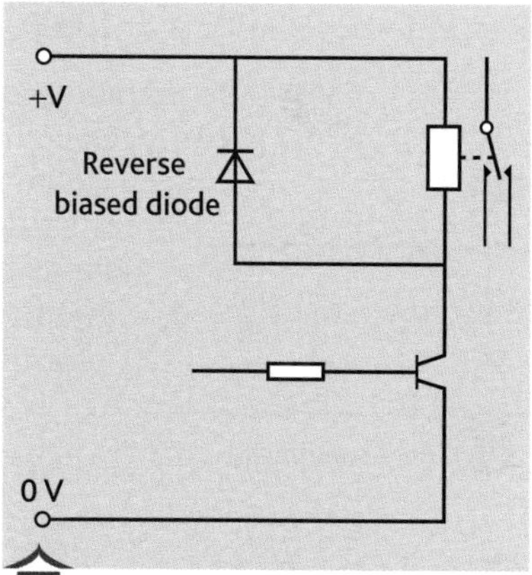

D *A circuit containing a clamping diode*

AQA Examiner's tip

- You should be able to identify the anode and cathode sides of diodes.
- Do not confuse the circuit symbol for a diode with that for an LED or thyristor.

kerboodle!

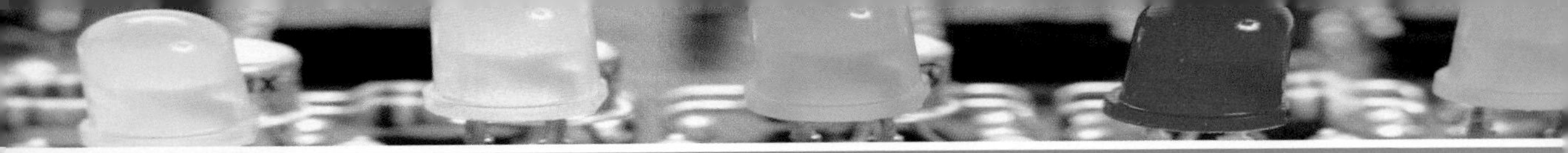

1.6 Resistors 1

Resistors are used to restrict or limit the flow of current in an electrical circuit. They are the most common component in electronic circuits and are used within input, process and output systems blocks.

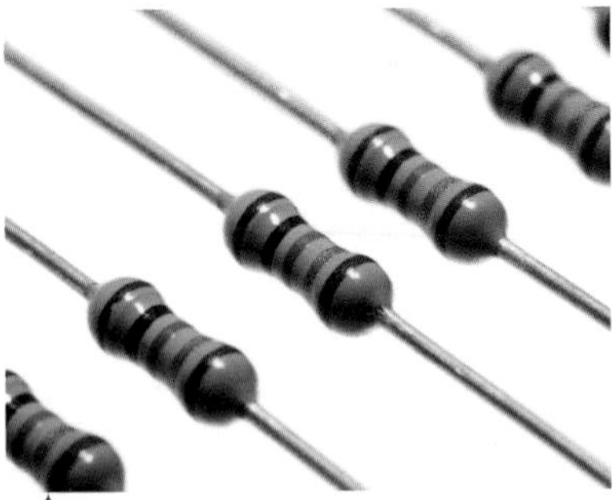

A *Examples of resistors*

Objectives

- Use Ohm's Law.
- Explain that resistors are used to control voltage and current in a circuit.
- Explain that LDRs are a type of variable resistor, where the resistance is changed by light.
- Explain that thermistors are a type of variable resistor, where the resistance is changed by temperature.

Key terms

Resistor: a component that limits current.

Ohm: unit of measurement of resistance.

Amp: unit of measurement of current.

Thermistor: a resistor where the resistance changes with temperature.

Calculating resistance: Ohm's Law

The resistance of a component is referred to by the letter R. The unit of measurement for resistance is the **ohm** (symbol Ω).

The resistance of a resistor can be worked out using the equation: **R = V/I**, where V is the voltage across the resistor and I is the current through it, measured in **amps**. For example, if the voltage drop across a resistor is 2 V and the current in the circuit is 20 mA,
$R = V/I = 2\ \text{V}/0.020\ \text{A} = 100\ \Omega$

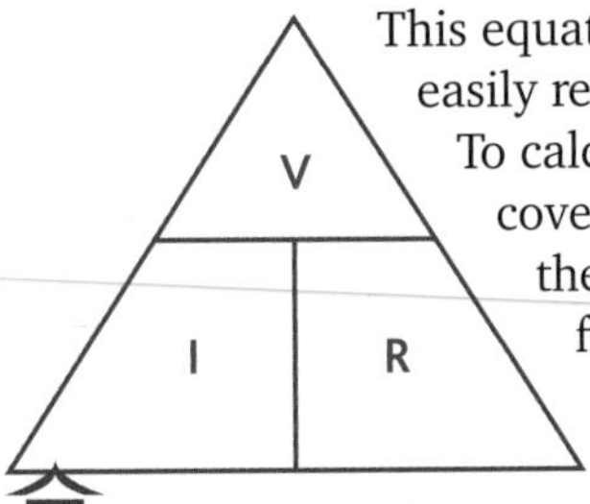

B *Ohm's Law triangle*

This equation comes from Ohm's Law. This can be easily remembered as a magic triangle, Figure **B**. To calculate any one of the three values R, V or I, cover up the letter that you want to know and then read off the formula. For example, to find the current, cover the letter I and the formula is V divided by R.

Remember

Current values

An **AMP** is really quite a large unit of current. When it comes to electronics, the milliamp is more typical. 1 milliamp (mA) is equal to a thousandth of an amp (A).

Types of resistor

Resistors are available in fixed values, covered in topic 1.7, or as variable resistors, covered here. There are also special types of variable resistor where the resistance can be changed by external conditions such as light and temperature, explained below.

Variable resistors

Variable resistors have three connections, one at each end of a resistance track and one moveable contact or wiper. The position of the wiper can be adjusted, meaning the length of track through which the current flows can be controlled. The variable resistor can be set up to make use of all three legs or just two legs, Diagram **C**.

If both end connections are used, the variable resistor acts as a voltage controller.	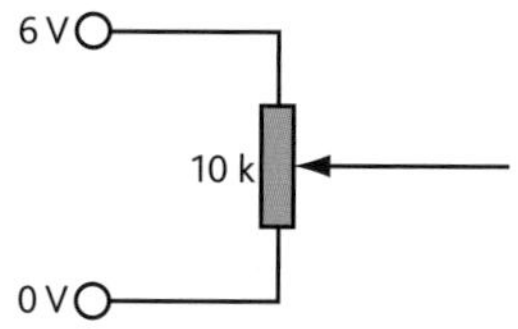
If only one end connection is used, the variable resistor acts as a current controller.	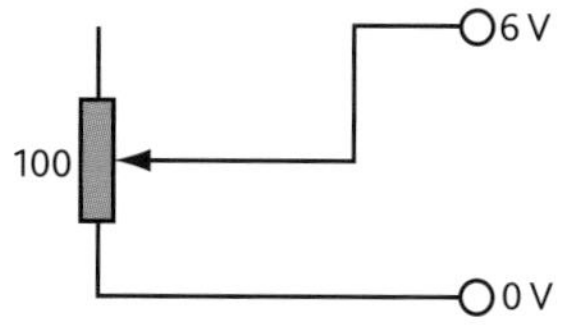

C *Variable resistor connections*

LDRs and thermistors

Light-dependent resistors (LDRs) and **thermistors** are special types of variable resistor. They are both made from semiconductor materials. They are often used as inputs to a system as part of a potential divider, which will be explained in topic 2.5.

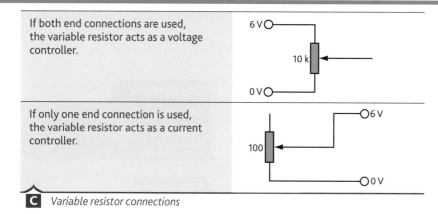

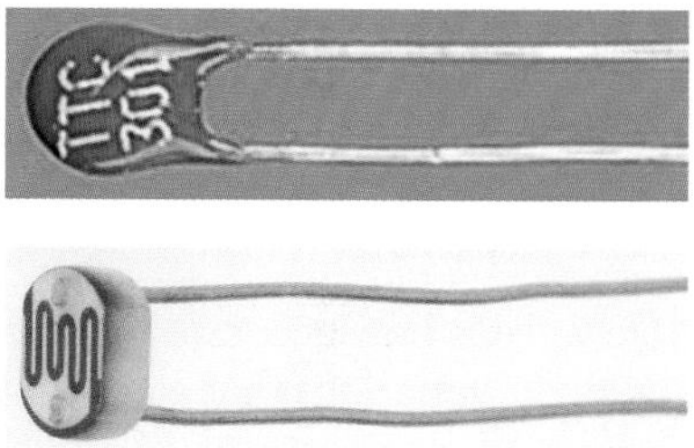

D *Examples of thermistors and LDRs*

Light-dependent resistor

The resistance of an LDR decreases as the amount of light that it is exposed to increases. In total darkness, the resistance of the LDR is very high, for example between 1 MΩ and 10 MΩ for some commercial products. In bright light the resistance can be very small, for example around 150 Ω for some commercial products. LDRs are used in applications such as security lights, where they save power by allowing the light to switch on only if it is dark.

Thermistor

For most common types of thermistor, as the temperature rises their resistance decreases. They are called negative temperature coefficient, or ntc, thermistors. Thermistors are commonly used for temperature measurement, switching and control applications.

Activity

Use a multimeter to measure the resistance of a thermistor. Now warm the thermistor between your fingers and note what happens to the resistance. Change the thermistor for an LDR. Note what happens to the resistance when it is exposed to light and when it is covered with your hand. Check your findings against the manufacturer's details for these parts.

AQA *Examiner's tip*

- You should be able to calculate the value of a current limiting resistor using Ohm's Law.
- You should be able to handle conversions between ohms, kilohms and megohms and between amps and milliamps.

Summary

Ohm's Law can be used to calculate voltage, current and resistance.

Fixed and variable resistors are used in circuits to control voltages and currents.

LDRs are a type of variable resistor that can be used to detect changes in light.

Thermistors are a type of variable resistor that can be used to detect changes in temperature.

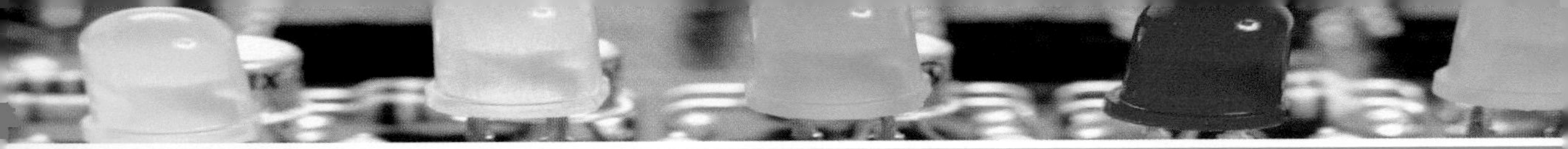

1.7 Resistors 2

Fixed value resistors

Fixed value resistors are most often used to protect components from excessive currents. They come in **preferred values**, as shown in the text box.

Colour code

All fixed resistors are marked with a colour code which shows the resistor's value in ohms, Table **A**. The colour code has four bands:

- Band 1 – the first number of the value
- Band 2 – the second number of the value
- Band 3 – the number of zeros after bands 1 and 2
- Band 4 – the **tolerance**. This band is normally gold (±5%) or silver (±10%).

Tolerance is how accurate the resistor is to its chosen value. This allows for slight variations that can occur when making resistors.

To read a colour code, the tolerance band should be placed on the right. For example, a resistor with the following bands: orange, orange, brown, silver would be 330 Ω ±10%. The tolerance means that it could be anywhere between 297 Ω and 363 Ω.

Colour	Digit
Black	0
Brown	1
Red	2
Orange	3
Yellow	4
Green	5
Blue	6
Violet	7
Grey	8
White	9

A *Resistor colour code table*

Objectives

Determine the value of fixed resistors and combinations of resistors.

Key terms

Preferred values: commonly available resistor values.

Tolerance: the possible variation in the accuracy of a resistor's or capacitor's value.

Remember

Standard values of fixed resistors

The **E24** range of resistors has the following values as standard:

10 Ω, 11 Ω, 12 Ω, 13 Ω, 15 Ω, 16 Ω, 18 Ω, 20 Ω, 22 Ω, 24 Ω, 27 Ω, 30 Ω, 33 Ω, 36 Ω, 39 Ω, 43 Ω, 47 Ω, 51 Ω, 56 Ω, 62 Ω, 68 Ω, 75 Ω, 82 Ω, 91 Ω and all the decades up to 1 MΩ. So, for example, resistors can be bought with values of 12 Ω, 120 Ω, 1.2 KΩ, 12 KΩ, 120 KΩ and 1.2 MΩ.

The value of a 100 ohm resistor can be written as either 100 Ω or, on a circuit diagram, 100R.

For the high value resistors we use a prefix instead of writing down all the zeros. Adding kilo (K) means multiply by a thousand; adding mega (M) means multiply by a million. This makes 1000 Ω = 1 kilohm (or 1 KΩ) and 1 000 000 Ω = 1 megohm (or 1 MΩ).

It is accepted practice to replace the decimal point with the multiplier letter used as the prefix. For example, 4700 ohms is 4.7 kilohms which is written as 4 k7. Similarly, 68 000 ohms is written as 68 k and 1 500 000 ohms is written as 1 M5.

Resistors in series

When designing circuits, a safe option is to choose the nearest preferred value which is higher than the calculated need. However, it is possible to make a desired resistance value from a combination of resistors of other values by connecting two or more resistors in series, Diagram **B**.

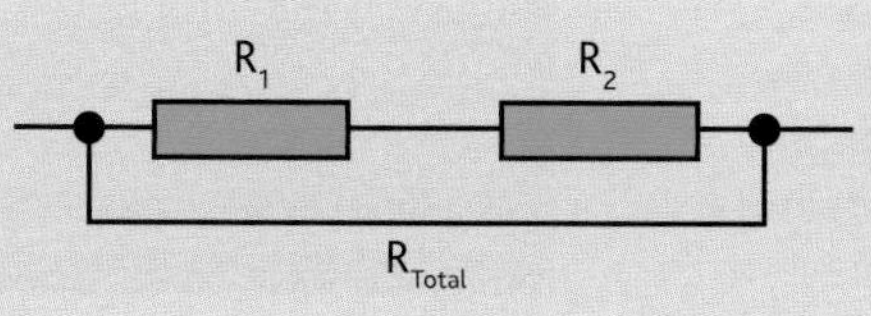

B *Resistors in series*

When resistors are joined in series the same current passes through each resistor. The total resistance, R_{Total}, is the sum of their individual resistances:

$R_{TOTAL} = R_1 + R_2$

For example, if a 51 Ω and 12 Ω resistor were put in series:

$R_{TOTAL} = R_1 + R_2 = 51 + 12\ \Omega = 63\ \Omega$

Resistors in series should not be confused with single in-line (SIL) or dual in-line (DIL) resistors. These are just different ways of packaging individual resistors so that they fit more easily onto circuit boards.

Resistors in parallel

When two or more resistors are connected in parallel, the voltage across each resistor is the same but the current flowing through each resistor will be dependent upon the resistance of each individual resistor. The total resistance of the resistors in parallel will always be smaller than the smallest individual resistance.

The formula to calculate the total resistance, R_{Total}, of resistors in parallel is:

$1/R_{TOTAL} = 1/R_1 + 1/R_2$

For two resistors in parallel, this is the same as:

$R_{TOTAL} = (R_1 \times R_2)/(R_1 + R_2)$

For example, if a 1 KΩ and a 100 Ω resistor were put in series:

$1/R_{TOTAL} = 1/R_1 + 1/R_2 = 1/1000 + 1/100\ \Omega = 11/1000\ \Omega$

Therefore $R_{TOTAL} = 1000/11\ \Omega = 90.9\ \Omega$

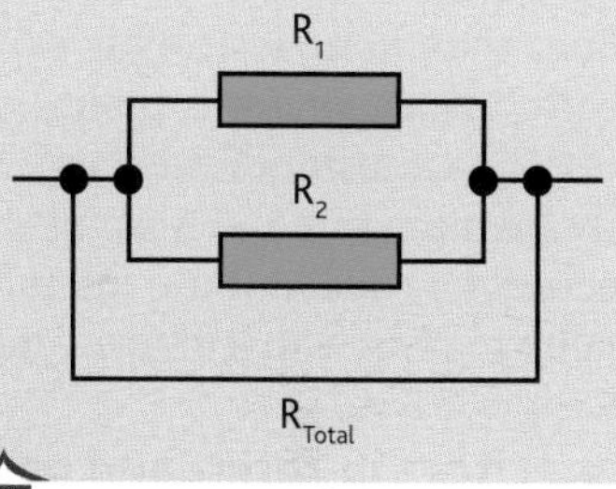

C *Resistors in parallel*

Activity

Work out the following resistor values from colours to numbers or numbers to colours:

- brown black brown silver
- yellow violet red gold
- brown red orange silver
- 10 kΩ ±10%
- 1.2 MΩ ±5%

AQA Examiner's tip

You should be able to interpret a resistor value using the colour code.

Summary

Fixed resistors are most often used to protect components from excessive currents.

Fixed resistors are marked with a colour code and can be combined in series to make new values.

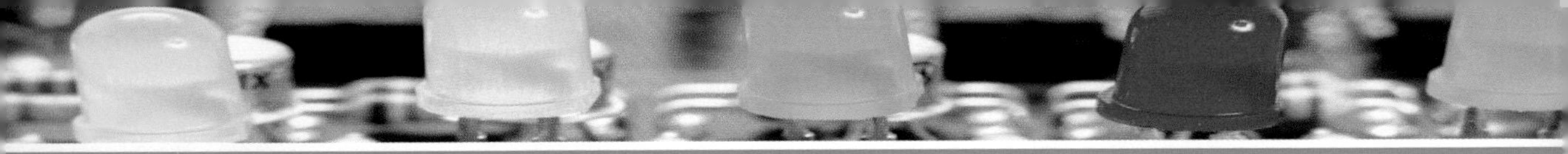

1.8 Capacitors

Capacitors

Capacitors are electrical components that are used to store charge. They are used in many different types of process block. When a capacitor is connected to a power supply it charges up, similar to a battery. If the capacitor is then disconnected from the power supply this charge can be used to push current through a circuit. For example, in Diagram **A**, when the switch is moved the LED will come on for a brief period as the capacitor discharges.

Types of capacitor

Capacitors are either electrolytic or non-electrolytic. **Electrolytic capacitors** are polarised components and must be inserted into the circuit the correct way round or they will not work. Non-electrolytic capacitors are not polarised and can be inserted either way round. Electrolytic capacitors can be either radial or axial and the direction of polarity is usually indicated on the case. Radial capacitors have different length legs. The short leg should be joined to the negative rail of the power supply.

Electrolytic capacitors can be manufactured with higher capacitance values than non-electrolytic capacitors. However, the tolerance of an electrolytic capacitor is typically around ±20%, but can be as high as ±50%. As a result of this large tolerance, electrolytic capacitors are only manufactured in multiples of 1, 2.2 and 4.7. Some types of non-electrolytic capacitor are manufactured to a tolerance of ±1%, which results in more capacitance values being available.

Capacitors in series and parallel

As electrolytic capacitors are only available in a limited range of multiples, it can sometimes be necessary to join capacitors together in series or parallel to make different values.

Series

When two or more capacitors are connected in series, Diagram **B**, the formula to calculate the total resistance, C_{TOTAL}, is:

$$1/C_{TOTAL} = 1/C_1 + 1/C_2$$

For example, if two 47 nF capacitors were put in series:

$$1/C_{TOTAL} = 1/C_1 + 1/C_2 = 1/47 + 1/47 \text{ nF} = 2/47 \text{ nF}$$

Therefore $C_{TOTAL} = 47/2 \text{ nF} = 23.5 \text{ nF}$

The total capacitance of the capacitors in series will always be smaller than the smallest individual capacitance.

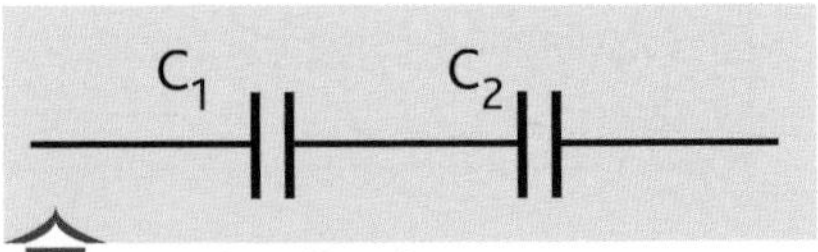

B *Capacitors in series*

Objectives

- Explain the difference between electrolytic and non-electrolytic capacitors.
- Describe how a capacitor can be used to produce a time delay.

Key terms

Capacitor: a component that stores charge.

Electrolytic capacitor: a capacitor that is polarised and will only work if attached the correct way round.

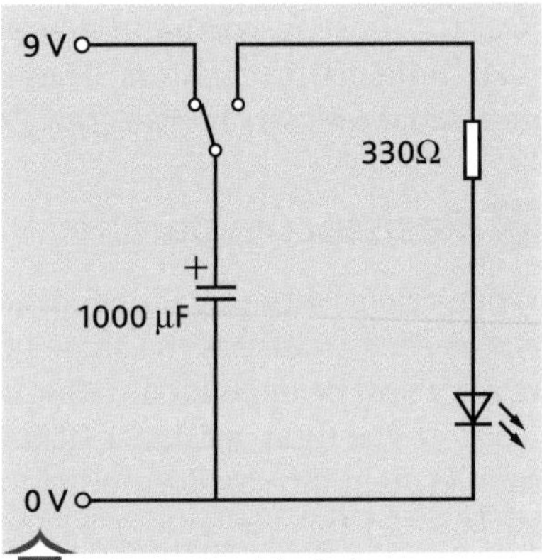

A *A simple circuit including an electrolytic capacitor*

AQA *Examiner's tip*

You should be able to identify different capacitors and explain the difference between them.

You should be able to calculate a time constant.

Parallel

When capacitors are joined in parallel, Diagram **C**, the total capacitance, C_{Total}, is the sum of their individual capacitances:

$C_{TOTAL} = C_1 + C_2$

For example, if a 47 nF and a 10 nF capacitor were put in parallel:

$C_{TOTAL} = C_1 + C_2 = 47 + 10\ \text{nF} = 57\ \text{nF}$

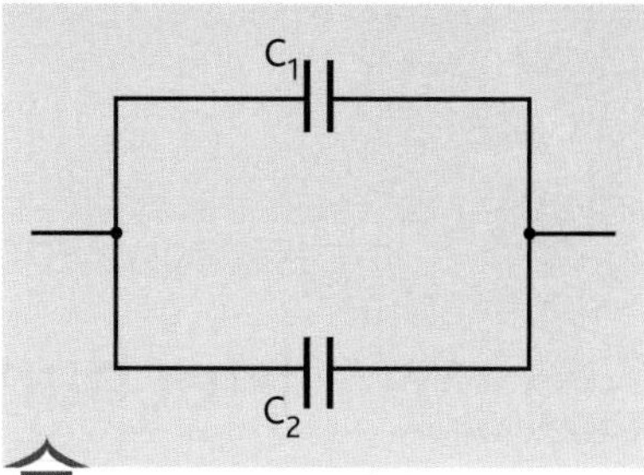

C *Capacitors in parallel*

Using capacitors in a timer circuit

Capacitors are often used in combination with resistors as part of a process block for a timer circuit. A capacitor can be charged instantaneously by connecting it directly to a battery. However, if a resistor is placed in series with the capacitor, Diagram **E**, this can be used to control the rate of charging. The time it takes for the capacitor to become charged depends upon the size of the capacitor and the value of the regulating resistor. This time delay is known as the **time constant** and can be calculated using the formula:

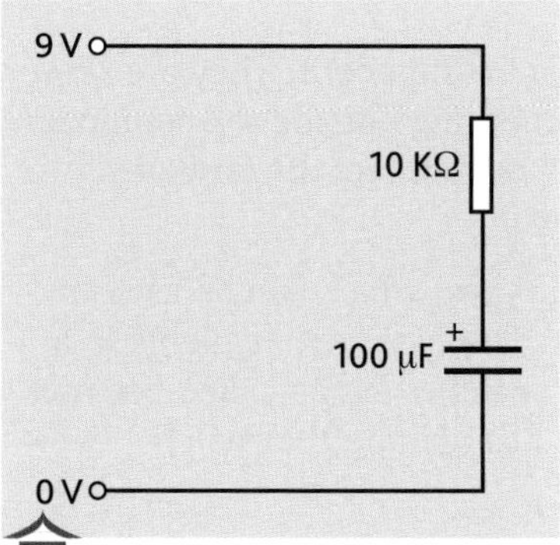

D *A capacitor and resistor in series*

Time constant (seconds) = Capacitance (farads) × Resistance (ohms)

For example, for the circuit shown in Diagram **D**

Time constant =
100 microfarads × 10 kΩ = 0.0001 farads × 10 000 Ω = 1 second

Another use of capacitors is to help to prevent damage to components due to power fluctuations within a circuit. Ceramic disc capacitors are often used to smooth an input across a supply, particularly when using DC motors.

Remember

Capacitance

The unit of capacitance is the farad. One farad is a very large quantity. Most capacitors used in schools are measured in microfarads (10^{-6} F, 0.000 001 F), nanofarads (10^{-9} F, 0.000 000 001 F) or picofarads (10^{-12} F, 0.000 000 000 001 F). The capacitance of a capacitor is normally printed on its case.

Activities

1 Calculate the time constant for a simple circuit including a capacitor with a value of 220 μF and a resistor with a value of 50 kΩ.

2 What values of capacitor and resistor could you use to achieve a time constant of 10 seconds?

Summary

Capacitors are used to store electricity. They may be either polarised electrolytic or non-polarised, non-electrolytic.

A capacitor can be used in series with a resistor to create a simple timer circuit.

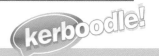

Remember

Supercapacitors

A growing use of capacitors is to replace conventional batteries or to provide supplementary storage of charge in a circuit. This uses supercapacitors, which will accept an unusually high charge compared to a normal capacitor. These components are highly efficient, do not degrade in the same way as conventional batteries and use less-toxic materials. However, they are currently more expensive than rechargeable batteries.

kerboodle!

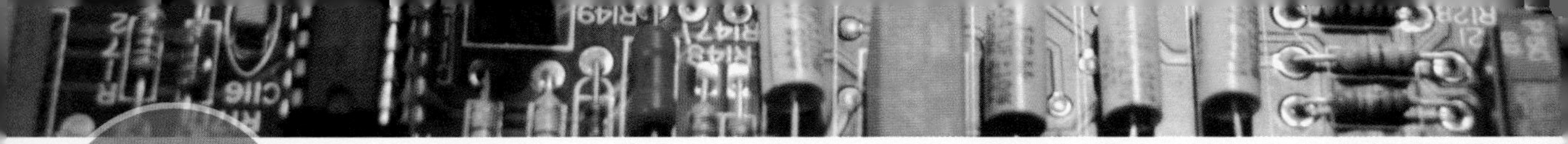

2 Inputs

2.1 Mechanical switches

Mechanical switches

Switches are the most basic type of input component. They are sensors, in that they have to be activated by something outside the component. They are used to turn circuits on or off, or to direct the flow of electricity along different parts of a circuit.

Switches come in a variety of different types, which can be used to perform different functions in a circuit. The types are available in a variety of different forms. Mechanical switches are so called because part of the component has to be moved to operate the switch.

Types of switch

Diagram **A** shows a timing circuit, using a 555 integrated circuit (IC). The two switches, SW1 and SW2, perform different jobs in the circuit. SW1 switches the circuit on and off. It is a single-pole, single-throw (**SPST**) switch. SW2 triggers the timing function of the circuit. It is a push-to-make (**PTM**) switch. The SPST switch **latches** whereas the PTM is **momentary**. This means that the SPST switch is either normally closed (**NC**) or normally open (**NO**) and only changes when someone (or something) turns it off. On the other hand, the PTM switch only stays on whilst it is being pushed.

Diagram **B** shows a different type of timer circuit. SW1 is again an SPST switch. It has two settings – on, when electricity can pass through it, or off, when it creates a break in the circuit. SW2 is a single-pole, double-throw switch (**SPDT**). Depending upon which way it is switched, it can direct the electricity around two different parts of the circuit.

Objectives

- List a number of different types and forms of switch.
- Explain that different types of switch perform different functions.
- Explain what switch bounce means and why mechanical switches may require debouncing.

Key terms

SPST: single-pole single-throw.

PTM: push-to-make.

Latches: stays on after being operated.

Momentary: switches off after being operated.

NC: normally closed.

NO: normally open.

SPDT: single-pole double-throw.

PTB: push-to-break.

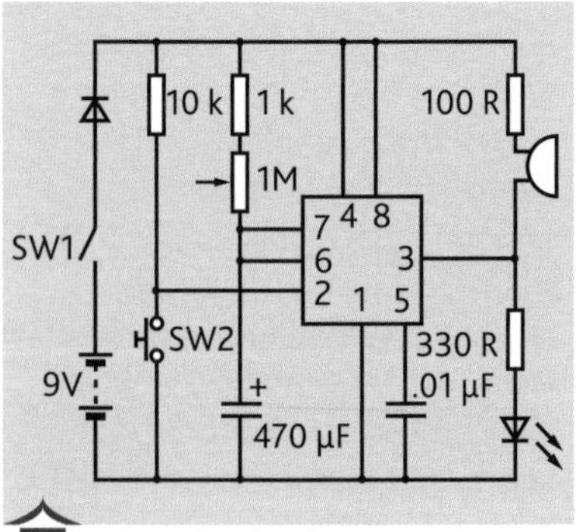

A *A timer circuit using a 555 IC*

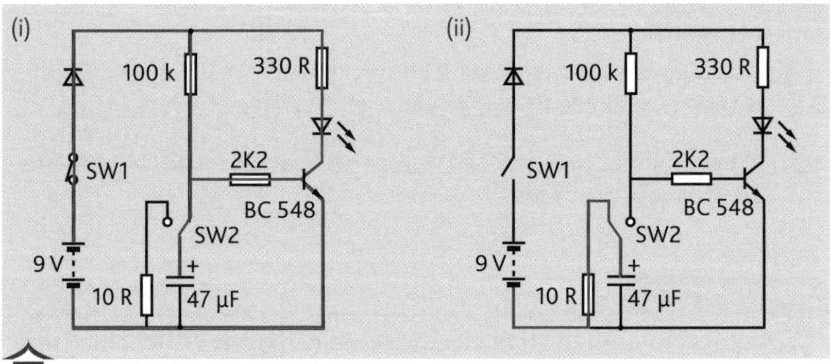

B *A timer circuit using a capacitor*

In operation, initially with SW1 closed, the electrolytic capacitor begins to charge. As the voltage in the capacitor rises, so does the voltage to the base leg of the transistor. When this reaches 0.7 V the transistor switches on and the LED lights, Diagram **B part i**. Changing the SPDT SW2 switch to the position shown in Diagram **B part ii** can discharge the capacitor. The 10R resistor is fitted so that the capacitor does not discharge instantly, which can cause arcing and damage the switch contacts. The circuit is now ready for another timing cycle.

Forms of switch

The switches in the circuits illustrated can be purchased in a number of different forms. For example, the SPST switch used for SW1 in both circuits (the on/off function) could be any one of the three shown in Diagram **C**, or a switch operated by a key. These are all latching switches.

Momentary switches, such as PTM and push-to-break (**PTB**) switches, are generally in the form of a button, Diagram **D**. One useful type of switch that usually has the capacity to be connected either as a PTM or as a PTB is the microswitch. These are often used as contact sensors on moving equipment and robot buggies. Another type of momentary switch is the reed switch. This is activated by movement, and is sometimes used as a motion sensor.

Debouncing mechanical switches

Mechanical switches contain contacts that move together and apart when the switch is operated. As they crash together they bounce apart slightly, sometimes several times before finally settling. This has the effect of producing a 'dirty' pulse instead of a clean digital pulse, Diagram **E**. It can sometimes be necessary to include additional components, such as a Monostable 555 IC (see topic 3.7) or a Schmitt trigger to stop this bouncing interfering with the operation of a circuit.

Activity

Switches are used on almost every electrical item in the home, ranging from the buttons on a remote control to the power switch on a kettle. Look around your home and identify at least three examples each of latching and momentary switches. State the items or places where you found them and explain why those types of switch were used in those applications.

Summary

Mechanical switches are available in a range of different types, such as SPST, SPDT, PTM and PTB.

The type of switch used will depend upon the job it is needed to perform.

The different types of switch are available in a wide range of forms. For example, latching SPST switches include rocker, slide and toggle types. Momentary PTM/PTB switches include buttons and microswitches.

Switch bounce happens where mechanical switches do not make and maintain a 'clean' contact at the first attempt. This can interfere with the operation of some electronic circuits.

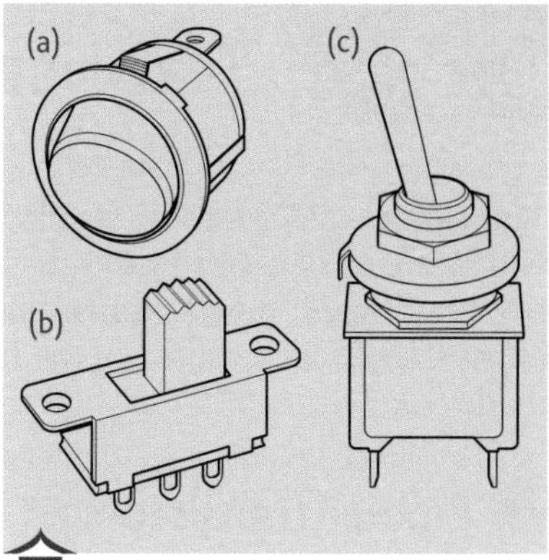

C *Examples of latching switches: (a) rocker switch, (b) slide switch, (c) toggle switch*

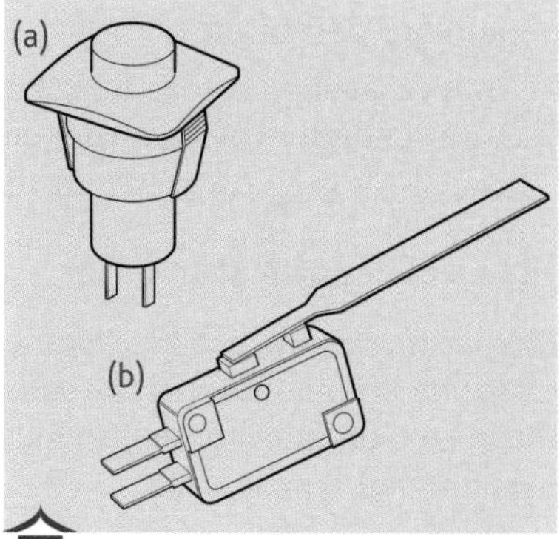

D *Examples of momentary switches: (a) PTB button, (b) microswitch*

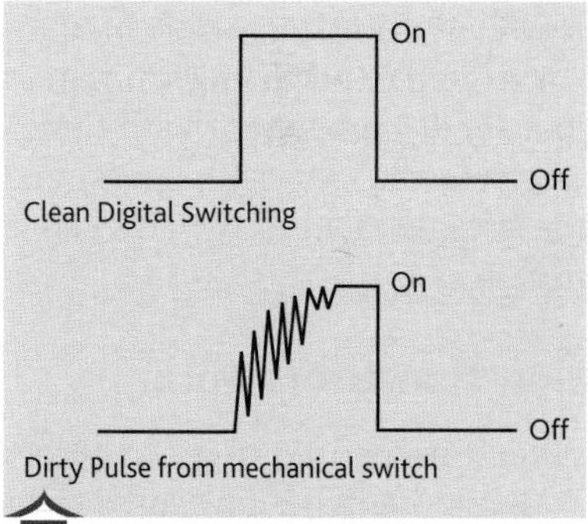

E *Effect of 'bouncing' on a mechanical switch*

AQA Examiner's tip

Be prepared to explain the functions of the PTM and SPST in the timer circuit.

kerboodle!

2.2 Transistors

Transistors are electronic switches and **amplifiers**. They are used in a wide range of systems blocks, including inputs and processes. There are two main types: bipolar transistors, investigated in this topic, and field effect transistors, covered in the next topic. Both types of transistor are made from layers of n-type and p-type semiconductor materials.

Objectives

Explain the operation of transitors as electronic switches.

Describe common uses of bipolar transitors.

Key terms

Transistor: a component that functions as an electronic switch and amplifier.

Amplifier: a device that can increase the output in proportion to the input.

Darlington pair: a combination of two transistors, normally in the same case.

Bias: the voltage (0.6 V) required to allow the transistor to be switched.

Gain: the amount of amplification provided by a transistor.

Bipolar transistors

The main uses for transistors are:

- to act as electronic switches
- to sense a change of resistance in a sensor device and switch on another part of the circuit
- to receive signals sent from low current devices and use these to turn on high current output devices, such as motors. Here the transistor is being used as a transducer.

Types of bipolar transistor

Bipolar transistors can be classified as NPN and PNP types, depending upon the arrangement of the semiconductor materials used to make them. In GCSE Electronic Products projects the most commonly encountered type is npn.

A simple transistor has three legs called the collector, base and emitter, Figure **A**. For an NPN transistor, the collector is the positive leg, the base is the input leg and the emitter is the negative leg. However, transistors can come in a variety of different case styles. For example, a **Darlington pair** is a combination of two transistors, in a single component with three legs, and an LM324 is a group of unconnected transistors packaged as an integrated circuit. Note that the arrangement of the legs can be different on different transistor packages.

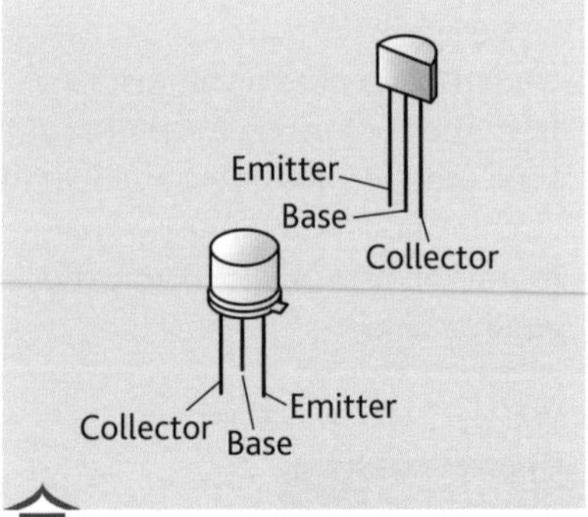

A *Examples of transistors*

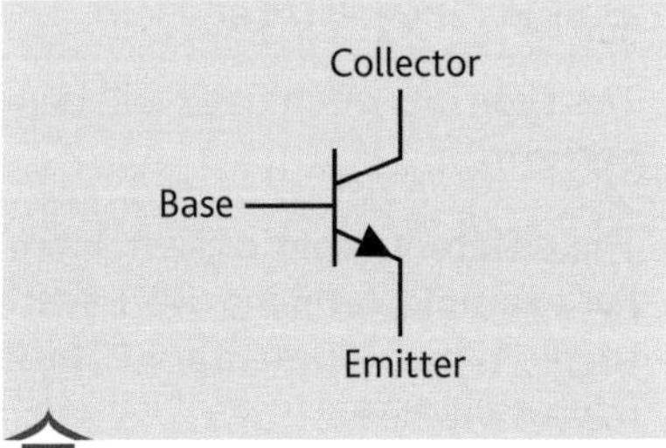

B *Circuit symbol for an NPN transistors*

How transistors work

When the base leg of a transistor receives a voltage of at least 0.6 V, it allows (switches on) some current flow from the collector leg to the emitter leg. This is sometimes known as **biasing** the transistor.

Transistors are analogue devices. As the base current increases, this allows a larger current to flow from the collector to the emitter. In this way, the transistor can be used to amplify the signal received at the base leg. A voltage of about 1.5 V between the base and the emitter turns the transistor fully on.

The amount of amplification is known as the **gain**. This is represented by the symbol ^{h}FE and calculated by dividing the current at the collector leg by the current at the base leg:

$$^{h}FE = I_C/I_B$$

For example, considering a BC548 transistor with a collector current of 100 mA and a base current of 0.5 mA:

$^{h}FE = I_C/I_B = 100\text{ mA}/0.5\text{ mA} = 200$

A common method of connecting two transistors together is called a Darlington pair or Darlington driver. The total gain of a Darlington pair is found by multiplying the gains of the transistors it includes:

$^{h}FE_{TOTAL} = {}^{h}FE_1 \times {}^{h}FE_2$

For example, for a Darlington pair where $^{h}FE_1 = 200$ and $^{h}FE_2 = 40$:

$^{h}FE_{TOTAL} = {}^{h}FE_1 \times {}^{h}FE_2 = 200 \times 40 = 800$

Transistor circuits as sensors

Diagram **C** shows a moisture detector circuit which includes a single transistor. When the resistance of the input sensor drops due to the electricity being conducted by the moisture, the voltage between the base and emitter will increase until it exceeds 0.6 V, when the transistor switches on. This will allow the current to flow through the resistor and light the output LED. This arrangement of the sensor and variable resistor is called a potential divider, and will be explained in 2.4.

The performance of this circuit can be improved by adding a second transistor and forming a Darlington pair, Diagram **D**. This provides a larger maximum collector current, which allows it to drive electronic devices that require more power, such as a buzzer.

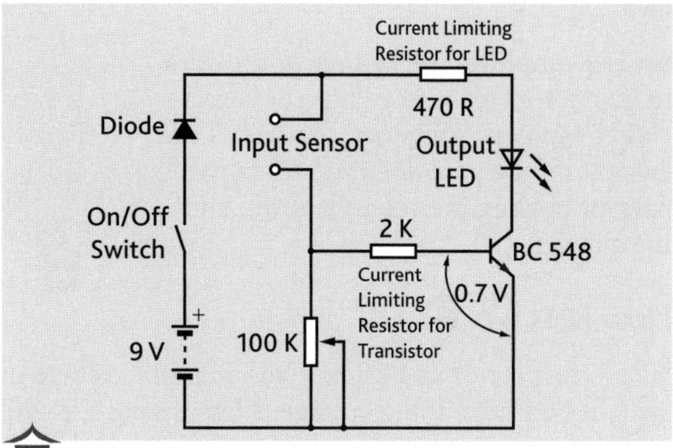

C *Simple moisture sensor circuit*

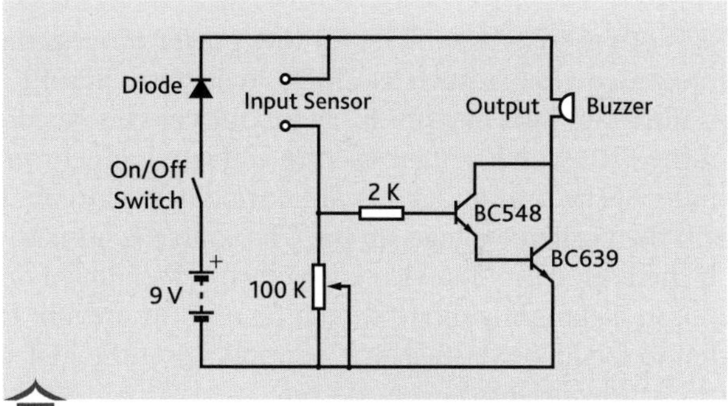

D *Moisture sensor circuit with warning buzzer*

Activity

This activity can be carried out using either computer software that simulates the operation of circuits or a breadboard model of a circuit. Create the circuit shown in Diagram C. Using a multimeter, measure the voltage differences between the three legs of the transistor when the output LED is off and when the output LED is on.

Summary

Transistors are electronic switches. They are analogue components that can be used to amplify current.

Common functions of transistors are electronic switching, sensing changes of resistance and driving high current outputs.

Different transistors can be used in combination (for example as a Darlington pair) to combine high amplification with high current load.

AQA Examiner's tip

- When building a circuit with a transistor, you need to protect the base of the transistor with a resistor to protect it from too much current.
- You need to protect a transistor from feedback (from electromechanical components such as motors) with a clamping diode.

kerboodle!

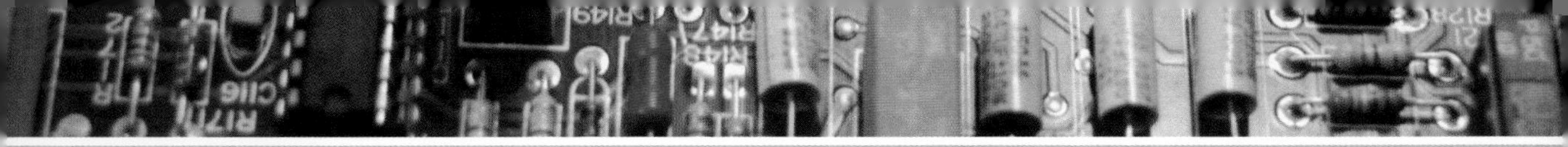

2.3 FETs

Field effect transistors

Similarly to bipolar transistors, field effect transistors (**FETs**) are made from a combination of n-type and p-type semiconductor material. They also have three legs. However, for an FET the legs are called the drain, gate and source, Diagram **A**. FETs come in a variety of different cases; for some designs, the metal case acts as one of the legs.

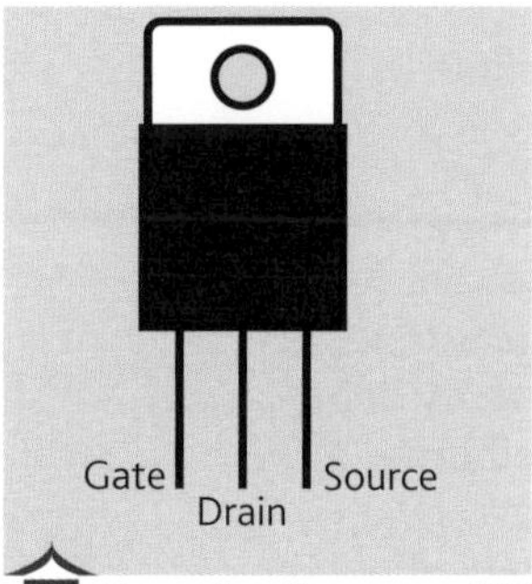

A *Example of legs on an FET*

An FET amplifies the voltage at the gate to gain an increase in voltage or current. Unlike bipolar transistors, the size of the current on the gate does not affect the current flowing between the drain and the source.

How FETs work

When the gate of an FET receives an input voltage of at least 2 V, it switches on fully. If the voltage is less than 2 V it will be fully switched off. This means that FETs are **digital** switches, whereas bipolar transistors have analogue capability.

FETs are used as amplifiers for low-power process units, in particular integrated circuits such as CMOS logic gates and PIC microcontrollers, as illustrated in Diagram **B**. These devices can be connected to the gate of the FET, which is then used to activate high current devices, such as motors. The output devices are normally positioned between the drain and the positive voltage supply. The source is normally connected to 0 V. If the high current device is an electromechanical device such as a motor, a clamping diode should be used to prevent the risk of damage due to feedback. In this type of application the FET is being used as a transducer driver.

FETs often have a metal backing or case to act as a **heat sink**, or to enable a bigger heat sink to be attached. This is because they can get hot when handling large currents. Some FETs contain circuitry to enable them to turn themselves off when they get hot and might get damaged, turning them into 'smart' transistors.

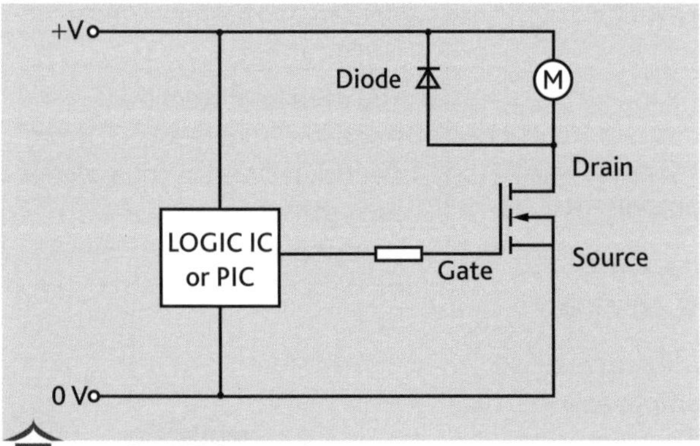

B *Use of an FET for interfacing*

Objectives

- Explain the operation of FETs as electronic switches controlled by voltage.
- Explain that FETs are used as transducer drivers and can supply high currents and switch high current devices on or off.
- Explain how FETs can be used as part of a sensor.

Key terms

FET: a field effect transistor.

Digital: of a signal that has only two states: high (on) or low (off).

Heat sink: a metal plate used to dissipate heat.

Impedance: resistance.

Using FETs as part of an Input

The gate of an FET has a high **impedance** and is not dependent upon current for switching. This means that FETs are more suitable for use with a touch pad operated by finger contact than a bipolar transistor, Diagram **C**. This is because the touch pad has high resistance and little current flows across fingers, but the voltage flow (minimum of 2 V) will be sufficient to switch the FET.

It should be noted that the supply voltage used in Diagram **C** is 6 V. This is because devices that use high current, such as motors, drain 9 V supplies almost instantaneously.

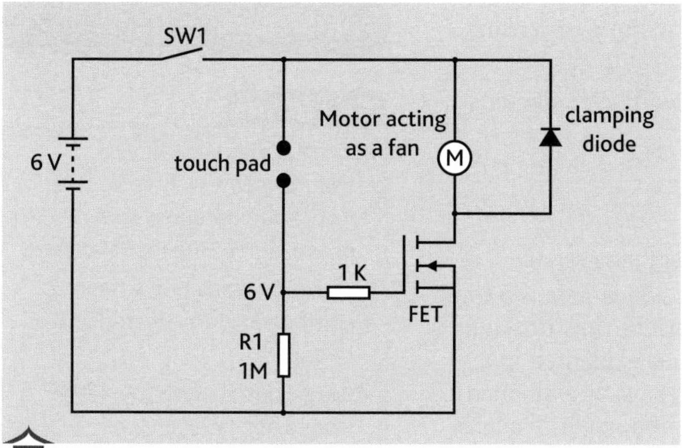

C *A touch sensor-operated fan circuit*

Activity

This activity can be carried out using either computer software that simulates the operation of circuits or a breadboard model of a circuit. Create the circuit shown in Diagram **C**. Using a multimeter, measure the voltage and current differences between the three legs of the FET when the touch pad is activated and when the touch pad is not activated.

Summary

FETs are electronic switches. They are digital components.

Common functions of FETs are as transducer drivers for high current outputs from low power process units such as integrated circuits and as electronic switches connected to high resistance/low current input sensors.

AQA Examiner's tip

- Remember the names of the legs of an FET.
- You need to protect an FET from feedback from electromechanical components such as motors with a clamping diode.

kerboodle!

2.4 Potential dividers 1

A **potential divider** can be used to divide the voltage in a circuit. They can be set up to provide either a constant or a variable input signal to the next part (or process block) of a circuit.

Potential dividers can be used in a number of different ways:

- to provide different voltages to different parts of a circuit from the same battery, explained in this section
- as part of an input block with suitable sensing devices
- to provide a process block called a comparator with these sensing devices.

The use of potential dividers with sensing devices will be explained in 2.5.

Objectives

Explain how a potential divider can be used to control voltages in a circuit.

Explain how a potential divider can provide a constant or variable input voltage to a process block.

Key terms

Potential divider: a device that divides a voltage so that its output voltage is some proportion of the input voltage.

Series: an orientation where components are located end to end.

Voltage signal: output voltage from a potential divider.

How a potential divider works

A potential divider is made up of resistors connected in **series**, Figure **A**. R_1 and R_2 are connected to a 9 V battery, which means a total of 9 V is dropped across the resistors. The same current runs through both resistors. The total resistance of the resistors, as explained in topic 1.8, is 300 Ω + 600 Ω = 900 Ω. From Ohm's Law, as explained in topic 1.7, voltage = current × resistance, so the voltage drop across each resistor is proportional to its resistance.

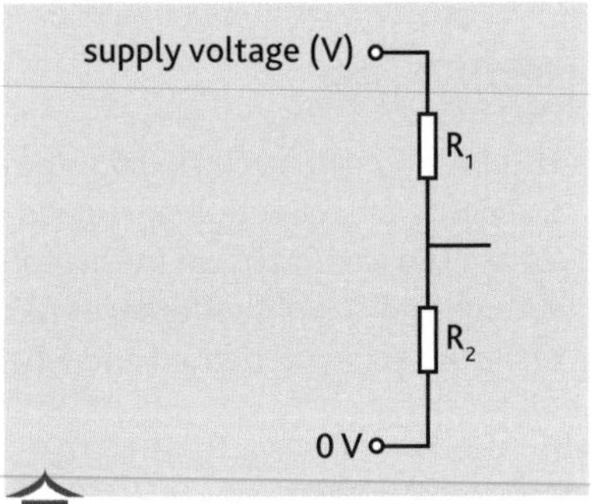

A *Potential divider*

The voltage across R_2 is called the **voltage signal** (V^s). This is the output from the potential divider and the input to the next part (or process block) of the circuit. For a given supply voltage, V, this can be calculated as follows:

$$V_s = \frac{R_2}{R_1 + R_2} \times V$$

Similarly, the voltage drop across R_1 can be calculated by substituting R_1 for R_2 in the top of this equation, or by subtracting V_s from V. For the example shown in Diagram **B**, using the equation gives:

$$\text{Voltage drop across } R_1 = 9\text{ V} \times \frac{300}{900} = 3\text{ V}$$

$$V_s\text{, Voltage drop across } R_2 = 9\text{ V} \times \frac{600}{900} = 6\text{ V}$$

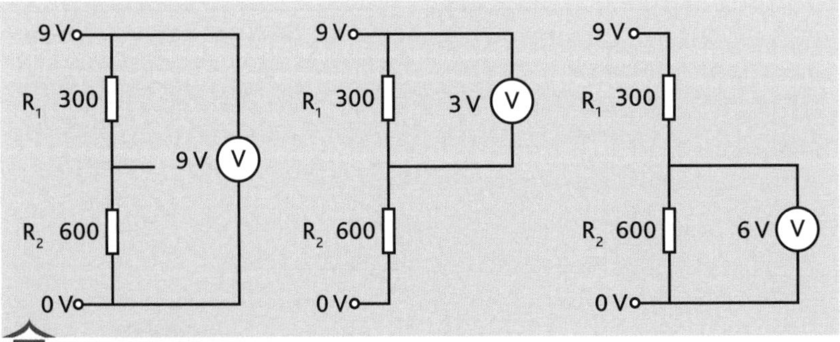

B *How a potential divider works*

Using potential dividers to divide voltage in circuits

Instead of using two separate fixed-value resistors as shown in Diagram **B**, a potential divider can be constructed using a single variable resistor such as a potentiometer, Diagram **C**.

If the third wiper leg on the potentiometer is set partway along its track, effectively the two different track areas on either side of the wiper operate in the same way as the separate resistors would. For example, if the third wiper leg is set halfway along its track (giving a resistance of 5 KΩ between points A and B and 5 KΩ between points B and C), it becomes equivalent to the circuit in Figure **C** using fixed value resistors. This would give an output voltage (V_s) of 4.5 V.

An advantage of using a potentiometer is that it allows for rapid and easy adjustment and a variable output.

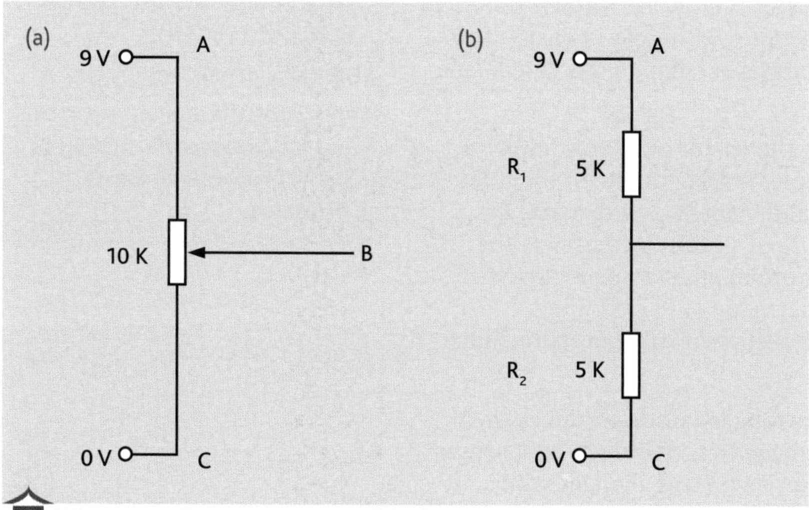

C *Using a potentiometer as a potential divider*

Remember

When using a variable resistor or potentiometer as one of the resistors in a potential divider, you should put a low value resistor (for example 1 KΩ) in series with it. This stops the full voltage being supplied as the voltage signal of the variable resistor is accidently set to zero ohms.

Activity

This activity can be carried out using either computer software that simulates the operation of circuits or a breadboard model of a circuit. Build a potential divider with R_1 = 10 and R_2 = 100 KΩ and connect them up to a 9 V supply. Use a multimeter to measure the voltage dropped across R_1 and R_2. What would happen if $R_1 = R_2$ = 100 KΩ? Try some other resistor values.

Summary

Potential dividers are created from resistors in series connected across a power supply.

The voltage signal is developed across the bottom resistor R_2.

Variable resistors can be used to create variable voltage signals.

AQA Examiner's tip

You will need to be able to calculate the voltage signal from a potential divider.

kerboodle!

2.5 Potential dividers 2

If one of the resistors in a potential divider is substituted with some form of **analogue** sensor, this can be used to provide a voltage signal at the output that varies with the intensity of the phenomenon being measured. Analogue sensors are those that monitor a phenomenon which is variable, such as light, sound or temperature.

Potential dividers are used with analogue sensors in a wide range of applications, such as the light sensor on security lights or automatic doors, temperature sensors in a heating system or a food cooling system and noise detectors in alarm systems.

Objectives

- Explain how a potential divider can be used with an analogue sensor.
- Explain how potential dividers can be used to compare a sensor to a reference value.

Key terms

Analogue: a signal that is variable (that is, does not just have two states of on or off).

LDR: light-dependent resistor.

Sensitivity: the amount a sensor's output changes with changes in the phenomenon being measured.

Sensing inputs using LDRs and thermistors

If one of the resistors in a potential divider is replaced with an **LDR**, an output voltage can be produced which varies with light intensity. This can be used to activate other process blocks or components in a circuit, such as transistors or thyristors.

Depending on the position of the LDR, the circuit can detect either a falling or a rising light level. For example, in Diagram **A** (a) the LDR is placed at the bottom on the potential divider (R_2) with a 100 KΩ resistor at the top (R_1). As the resistance of an LDR increases as the light level decreases, placing it at the bottom gives a rising resistance at R_2 as it gets darker. This results in the value of R_2 becoming significantly higher than R_1 and therefore gives a rising output voltage at V_s, creating a darkness sensor.

In contrast, Diagram **A** (b) shows the arrangement that would give a rising output voltage as the light level increases, creating a light sensor. You should note that the value of the series resistor changes with different arrangements.

Other sensors can be used with potential dividers in the same way, such as the thermistor setups shown in Diagram **B**.

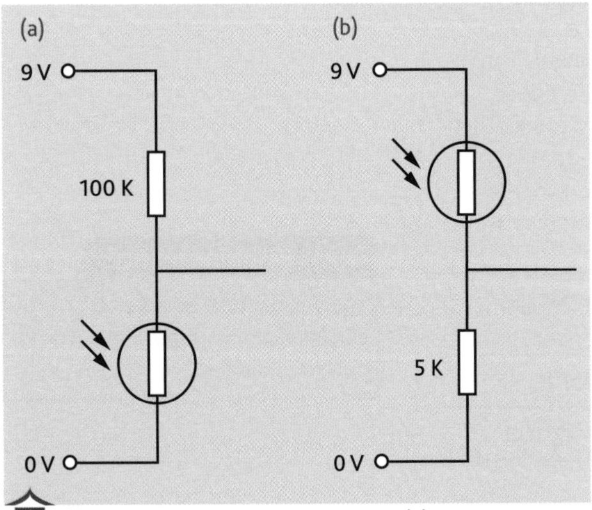

A *(a) voltage signal rises as it gets darker, (b) voltage signal rises as it gets lighter*

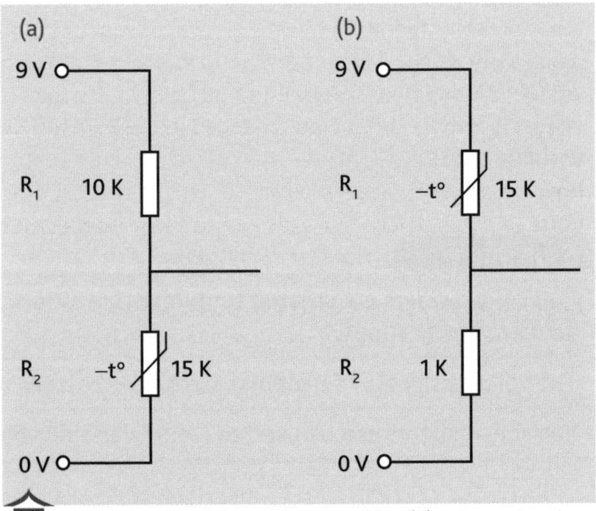

B *(a) voltage signal rises as it gets colder, (b) voltage signal rises as it gets warmer*

Using potential dividers as part of a comparator

A comparator process block compares two inputs. In one of its most common forms, it compares the input from a sensor against a target value. If the input is higher than the target value (or lower than the target value, depending upon the arrangement of the potential divider), this is used to trigger another part of the circuit.

Considering a potential divider with a thermistor and a resistor, if the fixed value resistor is replaced with a variable resistor, Diagram **C**, this allows adjustment of the **sensitivity** of the sensor. By selecting a suitable variable resistance, this means that the output signal can be designed to reach a certain 'trigger' value when the temperature reaches a certain level. This 'trigger' value could be enough to activate the next process block or component, such as a transistor (topics 2.2 and 2.3) or a thyristor (topic 3.1).

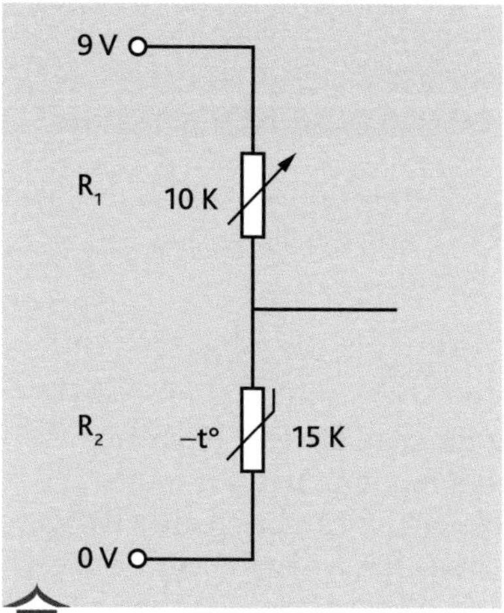

C *A variable resistor used to adjust the sensitivity of a thermistor*

The same principle applies when using other forms of analogue sensor, such as an LDR. However, as when using a fixed value resistor, you must note that the value of the variable resistor will be different depending upon the sensor used and the orientation (that is, whether the sensor is positioned at R_1 or R_2).

Remember

It is good practice to use a low value fixed resistor in series with the variable resistors as part of R1. See topic 2.4 for details.

Activities

1. An LDR set up as a 'darkness' sensor (Figure **A** (a)) has a resistance of 500 Ω in bright light and 200 KΩ in the shade. If R_1 = 10K, work out the values of V_s in bright light and in darkness.
2. Draw a potential divider that could be used as a sensor circuit in a fire alarm.

AQA Examiner's tip

You will need to be able to determine how the output from a potential divider is affected by swapping a sensing component between positions R_1 and R_2.

Summary

LDRs and thermistors can be used in a potential divider to create variable voltage signals from a potential divider. Depending upon how these components are arranged, they can be used to provide a voltage signal that increases as the sensor detects more or less light (or a colder or warmer temperature).

Replacing the series resistor used with an LDR or thermistor with a variable resistor can allow the sensitivity of the sensor to be controlled, allowing the potential divider to form part of a comparator.

3 Processes

3.1 Thyristors

Thyristors

A **thyristor** is an electronic component that can be used to create a latching circuit. It is sometimes called a flip-flop, but it is more correctly known as a **bistable**. A bistable circuit can be created using other electronic components, but in this section we will focus on the use of a thyristor, as this can be one of the cheapest ways of achieving this process block.

Similarly to transistors and FETs, thyristors have three legs. However, for a thyristor the legs are called the cathode, anode and gate, Figure **A**.

Thyristors are particularly useful in alarm systems, where somebody briefly breaks a light beam, or steps on a pressure pad. Even that short signal is sufficient to cause them to latch and make the alarm stay on until they are reset.

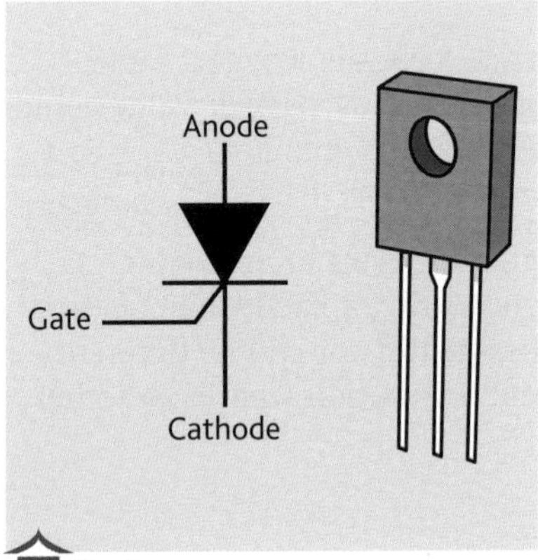

A *Thyristor*

How thyristors work

A bistable is a device which has two stable states: we can call these on and off. A thyristor will only allow current to flow from the anode to the cathode when it is turned on. It can be activated by a momentary signal, when a voltage of more than 2 V is applied to the gate. This needs only a very small current. Once the signal is removed the thyristor will remain on until it is reset (unlike a FET, which would turn off). Turning the thyristor off can be achieved by interrupting the current flow through it or by switching off the power supply completely.

Thyristors often have a metal backing or case to act as a heat sink, or to enable a bigger heat sink to be attached. This is because they can get hot when handling large currents.

Using thyristors as latches

Diagram **B** shows a circuit where the thyristor operates as a latch. When the PTM switch SW1 is pressed, the LED will turn on. It will remain lit until the PTM switch SW2 is pressed, as this will momentarily short-circuit the current that was flowing through the thyristor.

Objectives

- Explain how a thyristor is used to create a bistable circuit.
- Explain the operation of thyristors as latches.
- Explain how to switch a thyristor with a range of different input devices.

Key terms

Thyristor: a component often used as a bistable or latching device.

Bistable: a circuit which stays on after a momentary signal is received to the input. Also known as a latch or a flip-flop.

A significant advantage of thyristors is that the signal to the gate needs very little current. For example, piezoelectric material produces a small electrical charge when squeezed. If a small piezo transducer is pushed by a finger, the voltage created is sufficient to trigger the thyristor **C**. The buzzer sounds and the capacitor in parallel smoothes the supply. If a smoothing capacitor is not used, the pulsed signal of a buzzer can reset the thyristor. As for the previous example, the PTM switch SW2 acts as a reset for the alarm.

Similarly, a thyristor can be used in a light sensor, Diagram **D**. In this case, a potential divider is used to act as a comparator for the light level, as explained in topic 2.5. When the latching switch SW1 is closed, the circuit is activated. When the light level at the LDR reaches a value determined by setting the variable resistor, the LED and buzzer are activated until the reset switch SW2 is pressed.

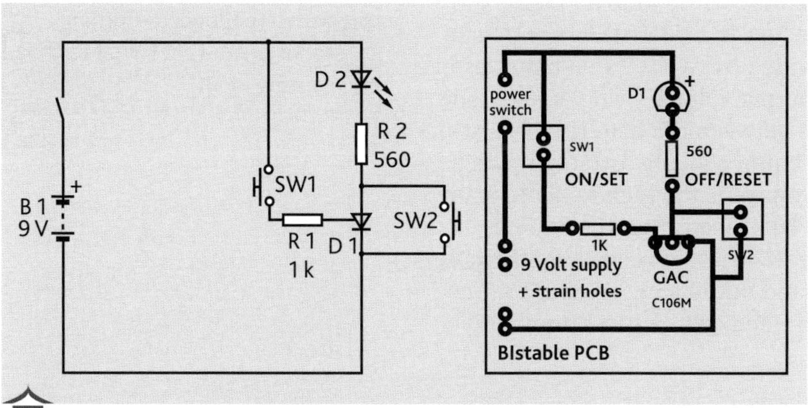

B *A latching LED circuit containing a thyristor*

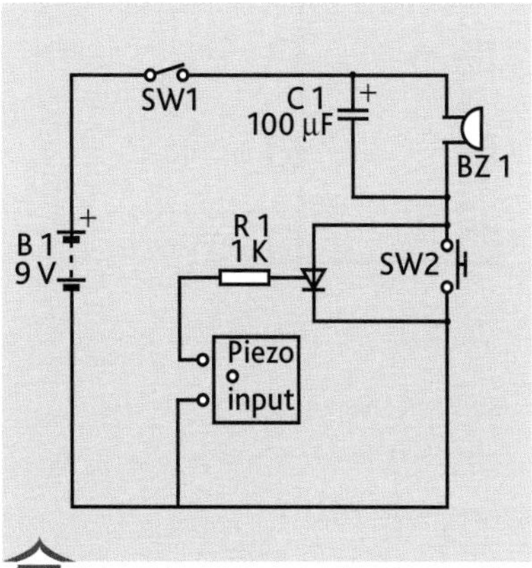

C *A latching alarm circuit triggered by a piezo contact*

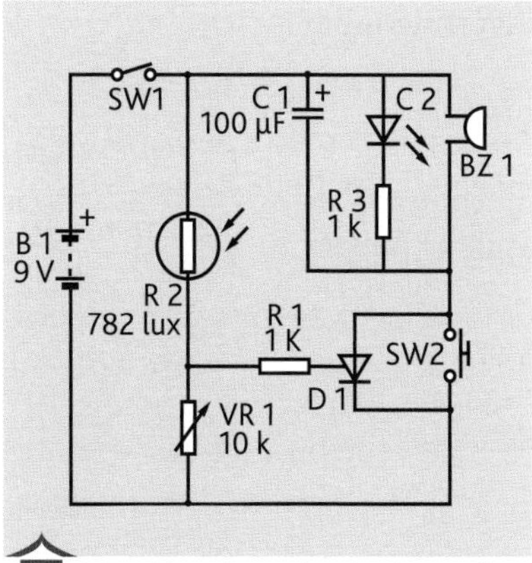

D *A latching alarm circuit triggered by a LDR*

Summary

A thyristor can be used to create a bistable circuit, which can function as a latch.

It can drive high current outputs.

The gate acts as the trigger. It is very sensitive and will switch on for even a momentary and very small electronic signal.

Activity

This activity can be carried out using either computer software that simulates the operation of circuits or a breadboard model of a circuit. Create the circuit shown in Diagram **B**. Using a multimeter, measure the voltage and current differences between the three legs of the thyristor before the switch SW1 is pressed, when SW1 is pressed and when SW1 is released.

AQA Examiner's tip

Remember the names of the legs on a thyristor and what they are connected to.

You should be able to explain how a thyristor can operate as a latch.

kerboodle!

3.2 Operational amplifiers – Op amps 1

The operational amplifier, or **op amp** to use its more common name, is a device used to amplify small differences between two input voltages. This difference is amplified to produce a voltage gain which can be as high as 100 000 – however, the output voltage cannot be any greater than the supply voltage. Op amps are used as comparator process blocks, to convert analogue signals to digital signals and to amplify voltage.

Objectives

Describe the function of an op amp.

Explain how to limit the gain of an op amp by using an input resistor and a feedback resistor.

Types of op amp

Op amps are integrated circuits (ICs), for example Diagram **A**. The two most common types of op amp are the 741 op amp and the 3140 FET op amp. The voltage output of the 741 is 2 V less than the power supply. In schools this is normally provided by two 9 V PP3 batteries in series, Diagram **C**, giving a maximum output voltage swing of +7 V and –7 V (0 V being the central connection between the batteries in series). It can be powered by a single 9 V PP3 battery, giving a voltage swing of +7 V and +2 V, but the minimum voltage is too high to allow many ICs and transistors to be switched off. This means that the 741 has been largely replaced in schools by the 3140, which costs slightly more but requires only a single PP3 battery and outputs to 9 V and 0 V. The 3140 is a direct replacement for the 741 and can be substituted into circuits designed for the 741.

Key terms

Op amp: operational amplifier.

Negative feedback: returning part of the output signal to the inverting input.

Normally on an eight-pin op amp IC, there is a dot marking the location of pin 1 and a notch showing which way round they should be placed. They have two inputs: the inverting input (marked –, pin 2 on the IC) and the non-inverting input (marked +, pin 3). They have one output (pin 6) and connections to the positive (pin 7) and 0 V/negative (pin 4) power supplies.

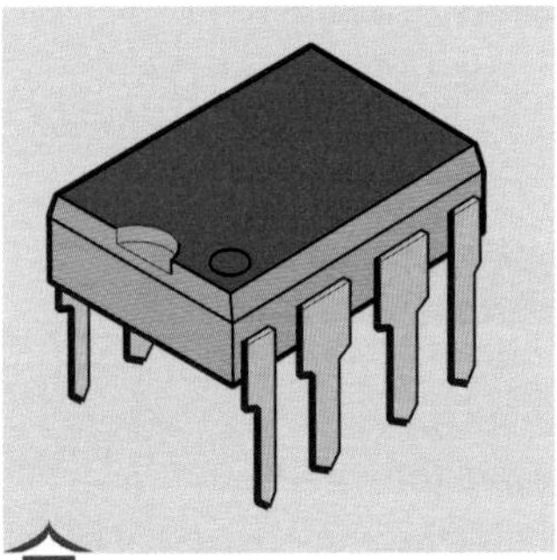

A *8 pin DIL IC*

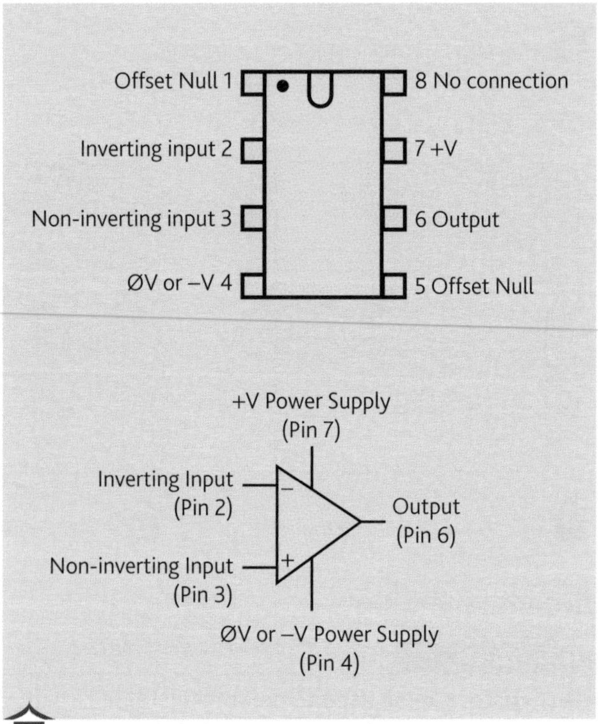

B *Circuit symbol of an OP Amp IC and pinouts*

Gain of an op amp

The gain of an op amp without any restrictions is about 100 000. This is called the open loop gain. This makes the op amp a highly sensitive device. It can detect the very small changes in voltage produced by the resistance of sensors varying marginally when being used in potential dividers. For example, if using a single 9 V power supply this represents a difference between the input pins of less than 0.000 1 V.

The gain can be restricted by using **negative feedback**. This requires a feedback resistor, R_f, and an input resistor, R_{in}, Diagram **D**.

The feedback resistor directs a small part of the output to the inverting input, ensuring that the feedback is negative. This makes the op amp more stable and the gain predictable. Gain is calculated using the following formula:

$$\text{Gain} = -R_f / R_{in}$$

For example, where $R_f = 50\ \text{K}\Omega$ and $R_{in} = 10\ \text{k}\Omega$,

$$\text{Gain} = -R_f/R_{in} = -50\,000/-10\,000 = -5$$

Gain has no units as it is a mathematical value. The negative sign indicates that the polarity of the output will always be inverted to that of the input. So if the input is a positive voltage, the output will be a negative voltage.

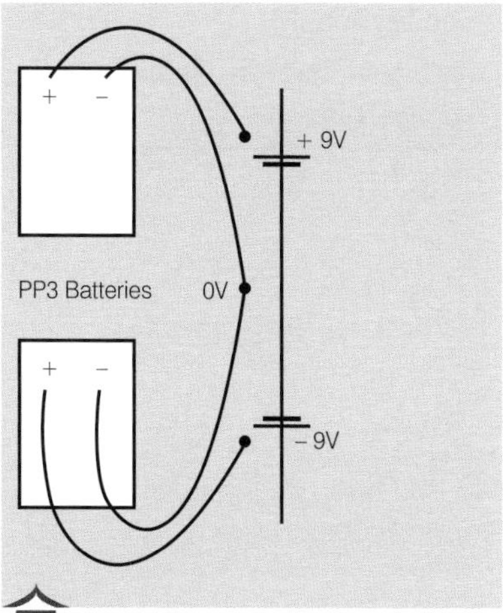

C *Use of dual power supply for a 740 op amp*

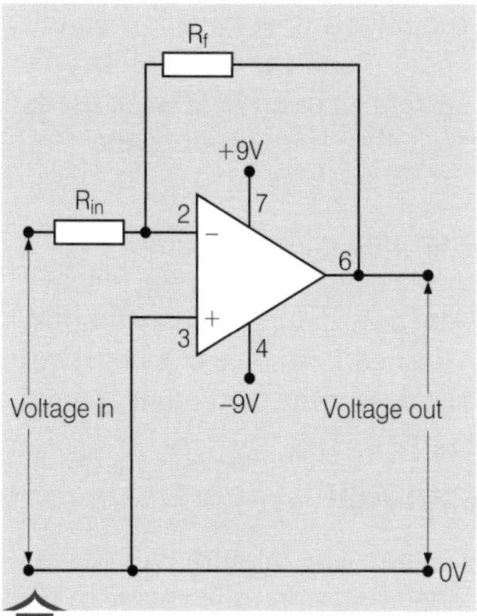

D *Use of a dual power supply for a 741 op amp*

AQA Examiner's tip

You should be able to identify the pins of an op amp and their functions.

Activity

Calculate the gain of an op amp with the following combinations of components:

- $R_f = 100\ \text{K}\Omega$, $R_{in} = 5\ \text{K}\Omega$
- Input resistor = 10 KΩ, feedback resistor = 1 MΩ

Summary

An op amp is an IC device used to amplify small differences between two input voltages.

The open loop gain of an op amp is 100 000. The gain can be limited by using an input resistor and a feedback resistor.

kerboodle!

3.3 Operational amplifiers – Op amps 2

How the op amp works as a comparator

A **comparator** process block compares two inputs. In one of its most common forms, it compares the input from a sensor against a target value. If the input is different from the target value, it provides an output that can trigger another process block or part of the circuit.

When the op amp is used as a comparator it compares the voltages of its inverting and non-inverting inputs. The op amp is able to detect very small differences between the two inputs. It multiplies these differences by its gain, which for open loop is about 100 000. In practice, this means that the output from the op amp is either the maximum or the minimum voltage of the system. If the non-inverting input is greater than the inverting input, the output is high voltage. If the inverting input is greater than the non-inverting input, the output is low voltage.

Diagram **A** shows a circuit that uses an op amp comparator to detect changes in light levels. The voltage at the non-inverting input is a **reference voltage**, being generated by a potential divider that includes a potentiometer. If the voltage at the non-inverting input (marked +) is higher than the inverting input (marked –), the output will be high and the **LED** will be on. Conversely, if the inverting voltage is greater than the non-inverting, the output will be 0 V and the LED will be off. If an output device requiring a higher current than an LED were used, this would need a transducer driver attached to the output of the op amp.

The combination of the high gain of the op amp with its digital output can lead to **clipping**. This is distortion of the signal as the input signal has been amplified beyond the value of the power supply. To avoid this, the gain of the op amp can be set to the desired level by using negative feedback, as explained in topic 3.2 on calculating the gain of an op amp.

How the op amp works as an ADC

The input to an op amp is an analogue signal. The op amp will check this signal many times every second, depending upon the gain – in open loop it may be as few as 10 times per second, but if the gain is limited to, for example, 1000 it will check the signal 1000 times per second. The output from using an open loop op amp has only two states (high or low), which means that it is digital. The combination of the high switching speed and the input and output states means that the op amp can be used as an analogue-to-digital converter (**ADC**). This means it is suitable for connecting analogue sensors such as LDRs and thermistors to logic circuits.

Objectives

- Explain how to use an op amp as a comparator.
- Explain how an op amp can be used as an ADC.
- Explain how an op amp can be used as an inverting amplifier.

Key terms

Comparator: a device that compares two values.

Reference voltage: a voltage value set by the user.

LED: light-emitting diode.

Clipping: distortion to the signal caused when it is amplified beyond the voltage of the power supply.

ADC: analogue-to-digital converter.

How the op amp works as an inverting amplifier

An inverting amplifier uses negative feedback, through a feedback resistor between the output and the inverting input. This was described in topic 3.2 explaining how to calculate the gain. It reads the input signal, amplifies it by the gain, and provides an output with the opposite polarity to the input, Diagram **B**.

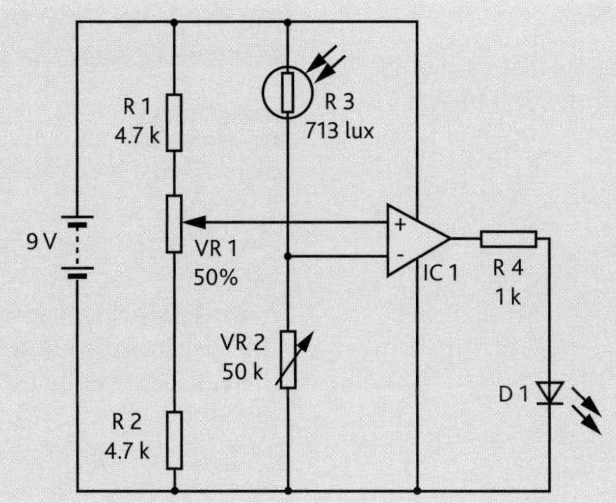

A *Input and output of an inverting amplifier*

Activity

This activity can be carried out using either computer software that simulates the operation of circuits or a breadboard model of a circuit. Create the circuit shown in Diagram **A**. Using a multimeter, measure the voltage differences between the two input legs of the op amp and between 0 V and the output, when the output LED is off and when the output LED is on. Measure the current passing through the LED and the op amp in both states.

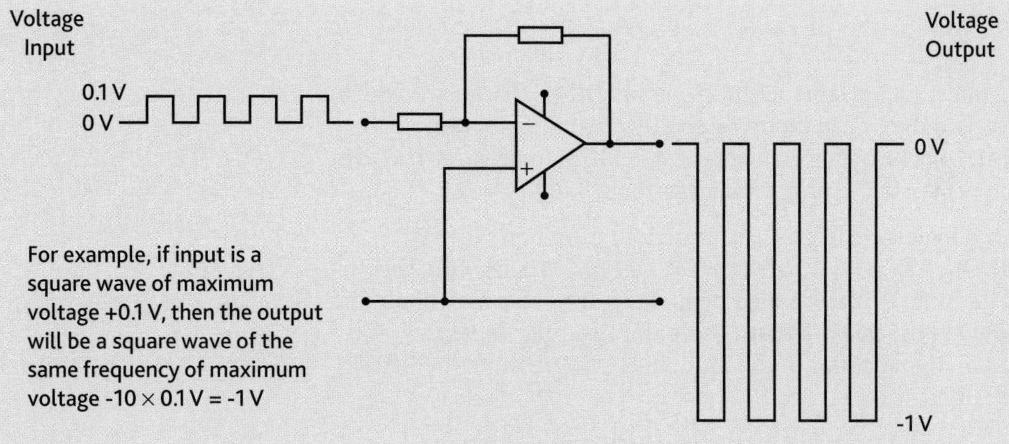

B *A light sensor using an op amp*

Summary

An op amp can be used as a comparator by linking a reference voltage from a potential divider to the non-inverting input and an analogue sensor to the output.

An op amp comparator can function as an ADC if the voltage gains exceed the available power supply, such as with open loop gain.

An inverting amplifier can be created by using an input resistor and negative feedback from the output to the inverting input through a feedback resistor.

AQA Examiner's tip

When using op amps for any output device with a higher current requirement than an LED you will need to use a transducer driver.

kerboodle!

3.4 Logic gates – OR gates

Logic gates

Logic gates are used where an electronic system needs to make a decision based on a number of inputs. They work according to strict rules and their outputs are determined by the inputs.

Logic gates are digital electronic devices. This means that each of the input and output gates must be in one of two states, Diagram **A**:

- **high**, also known as 1 or on
- **low**, also known as 0 or off.

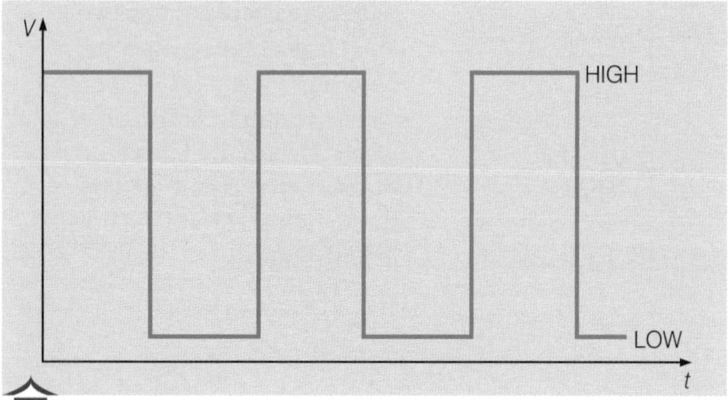

A *Digital signal*

Analogue signals, such as the output from an LDR or thermistor which can vary over a range, must be processed by an analogue-to-digital converter (ADC) before they can be processed by a logic gate. The use of an op amp as an ADC was explained in topic 3.3.

The name of a logic gate explains its function. Common types of gate include OR, AND, NOT, NAND, NOR and EXCLUSIVE OR. The output of a gate can be determined by looking at a table showing the possible input states, called a **truth table**, for example Table **C**. A and B represent the digital states of the input and Q represents the digital state of the output.

Each type of gate can be made from a combination of discrete components, but they are normally used as DIL integrated circuits (ICs). There may be a number of logic gates of the same type inside one IC package. They will operate off a common power supply to the IC but each gate will work separately. Although the ICs for different types of gate may look similar, on a circuit diagram they are represented by different symbols.

OR gates

The **OR gate** has two inputs. If either of these is high, the output is high, Table **C**. This can be illustrated by considering a mechanical OR gate, constructed from two switches in parallel, Diagram **D**. If either switch is pressed, or if both switches are pressed together, the light will illuminate.

Objectives

Explain what a logic gate is.

Describe the operation of an OR gate.

Key terms

High: a digital state also known as 1 or on.

Low: a digital state also known as 0 or off.

Truth table: a chart that explains the relationship between the inputs and the output of a logic gate.

OR gate: a logic gate that requires only one input to be high in order to produce an output.

Pinout: a diagram that shows what each pin does.

AQA *Examiner's tip*

You should be able to understand and use truth tables, know the logic symbols and be familiar with the pin layouts of a 14-pin DIL IC.

A common use of OR gates is in supermarket doors. We want the doors to open if someone wants to come in OR if someone wants to get out.

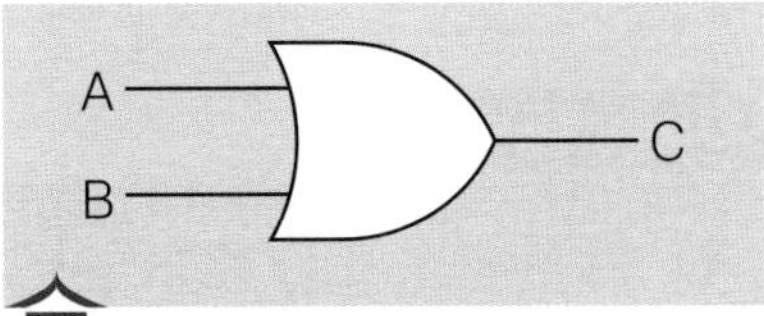

B *Symbol for an OR gate*

C *Truth table for an OR gate*

A	B	Q
0	0	0
0	1	1
1	0	1
1	1	1

Using OR gates in electronic circuits

The most commonly used OR gate in schools is the 4071 Quad Dual Input OR IC. Inside the chip there are four separate OR gates (QUAD), each with two separate inputs (DUAL) and an output. The **pinout** is shown in Diagram **E**. Diagram **F** shows a simple circuit that includes a 4071 IC. If either switch is pressed the LED will turn on.

The LED will also turn on if both switches are pressed at the same time. If a circuit were needed where the output would only go high if either switch were pressed on their own, this would use an EXCLUSIVE OR gate. If a circuit were needed that would operate only if all the inputs were low (that is, the opposite to an OR gate), this would use a NOR gate.

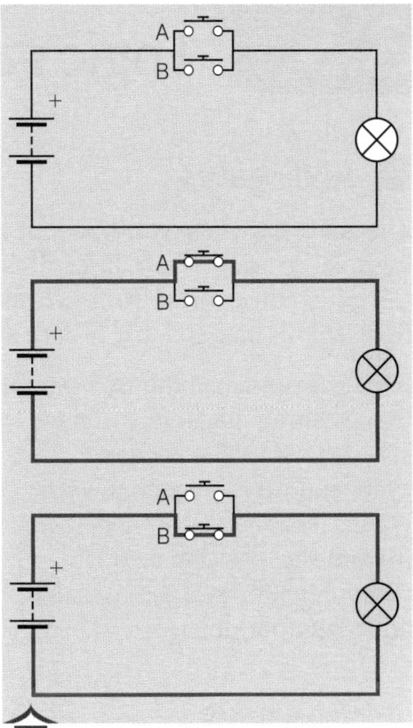

D *A mechanical OR gate: if either switch is pressed, the lamp will light*

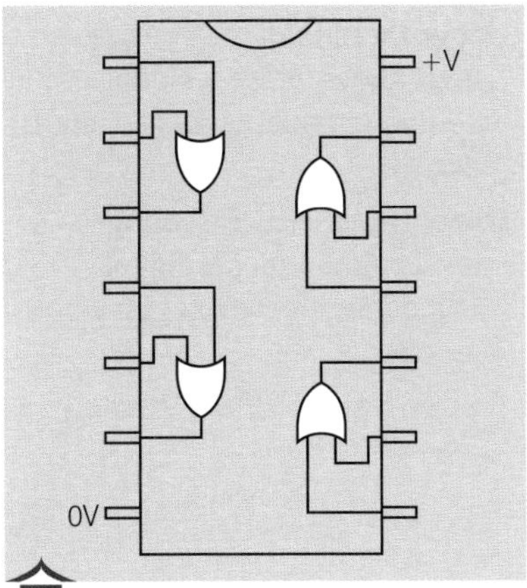

E *Pinouts for a 4071 IC*

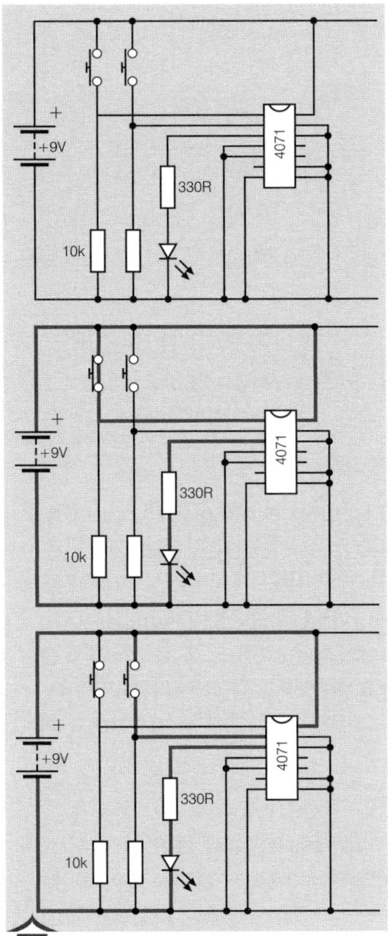

F *OR gate circuit: if either switch is pressed, the LED will light*

Activity

This activity can be carried out using either computer software that simulates the operation of circuits or a breadboard model of a circuit. Create the circuit shown in Diagram **F**. Using a multimeter, measure the voltage and current at the output of the IC for each of the possible combinations of switch operation, for example no switch pressed, each switch pressed individually, both switches pressed together.

Summary

A logic gate is a digital electronic device used to determine the state of an output based on a number of inputs.

An OR gate makes the output high when any of the inputs is high.

An OR gate can be constructed using the 4071 IC.

3.5 Logic gates – AND gates

AND gates

The **AND gate** has two inputs. If both of these are high, the output is high, Diagram **B**. This can be illustrated by considering a mechanical OR gate, constructed from two switches in parallel, Diagram **B**. The light will illuminate only if both switches are pressed at the same time.

AND gates are commonly used for safety applications. For example, in a washing machine you have to close the door AND press the start switch to start the washing cycle and fill the machine with water. If there was no AND gate it would be possible to start running water into the machine with the door open.

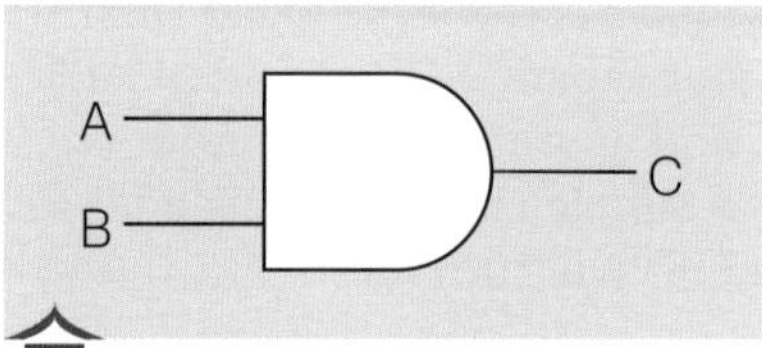

A *Symbol for an AND gate*

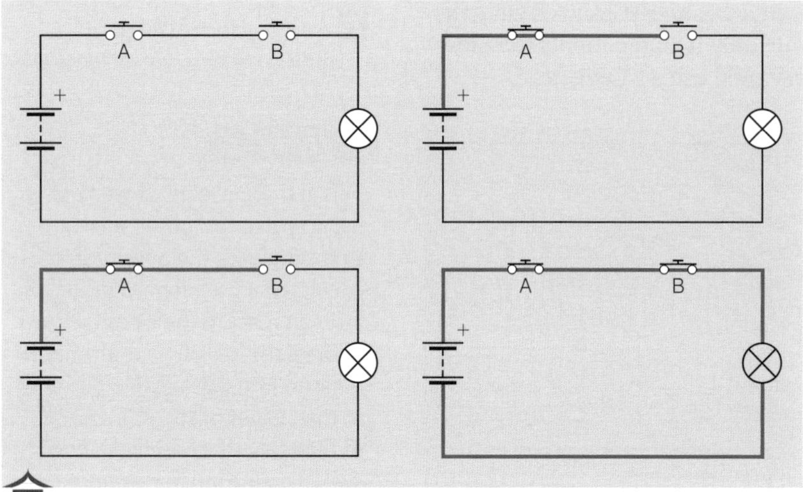

B *A mechanical AND gate: if either switch is pressed the lamp will light*

Objectives

Describe the operation of an AND gate.

Explain the need for pull down resistors with a logic gate IC.

Explain what is meant by the upper and lower threshold voltage for a digital component.

Key terms

AND gate: a logic gate that requires both inputs to be high to produce an output.

Pull down resistor: a resistor used to tie down inputs to prevent false readings due to static electricity.

Upper threshold: the voltage level above which a digital component recognises an input as high.

Lower threshold: the voltage level below which a digital component recognises an input as low.

Tie down: link to the 0 V supply.

Using AND gates in electronic circuits

The most commonly used AND gate in schools is the 4081 Quad Dual Input IC, Table **C**. The only visible difference between this IC and the 4071 OR gate is the number on the top. Diagram **D** shows the pinouts.

Diagram **E** shows a simple circuit using a 4081. The LED will only turn on if both switches are pressed at the same time. If a circuit were needed where the output would be high unless both switches were pressed at the same time (that is, the opposite to an AND gate), this would use an AND gate.

C *Truth table for an AND gate*

A	B	Q
0	0	0
0	1	0
1	0	0
1	1	1

Pull down resistors

The two 10 k resistors connected to the input pins in Diagram **F** are referred to as **pull down resistors.** Electronic logic circuits may not work properly without them.

Pull down resistors are needed because air, particularly on hot dry days, can charge objects with static electricity. This unwanted charge can cause the input pins to float high and make the IC think that an input is on. This would mean that the chip may switch itself on and off without warning. The pull down resistors literally pull the input pins down to 0 V so that the IC knows that the input is off. When the switch is operated the input receives the full supply voltage and then knows that it is on.

Threshold voltages

IC logic gates such as the 4081 are digital components. In practice, they have an **upper threshold** and a **lower threshold** to recognise whether an input is high or low.

The upper threshold is normally set at two-thirds of the supply voltage. The lower threshold is normally set at one-third of the supply voltage. For the circuit in Diagram **F**, which has a 9 V supply, any input over the upper threshold of 6 V would be registered as high. Any input below the lower threshold of 3 V would be registered as low. Any input voltage in between would have no effect on the current state of the IC.

It is good practice to **tie down** any unused inputs to 0 V.

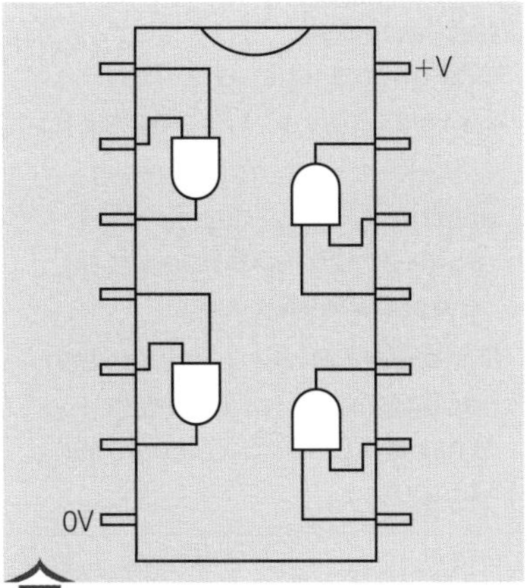

D *Pinouts for a 4081 IC*

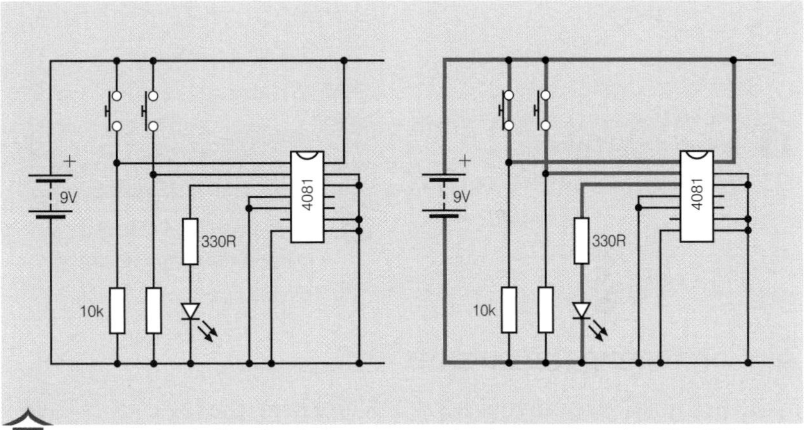

E *AND gate circuit: if both switches are pressed at the same time, the LED will light*

Activity

Make a list of all the AND gates that you can find in the school workshop and at home.

AQA Examiner's tip

You should use pull down resistors and tie down unused inputs to ensure that logic gate ICs operate as intended.

Summary

An AND gate makes the output high when both of the inputs are high.

Pull down resistors are needed with logic gate ICs to prevent static electricity causing pins to float high and register a false input.

Digital ICs have an upper threshold voltage above which they will recognise an input as high and a lower threshold voltage below which they will recognise an input as low. These are normally two-thirds and one-third of the supply voltage, respectively.

kerboodle!

3.6 Logic gates – NOT gates

NOT gates

The **NOT gate** is a digital device with one input. It is also known as an **inverter** because it quite literally inverts the input, Table **A**. If the input is high it will provide a low output. If the input is low it will provide a high output. A mechanical NOT gate can be constructed using a PTB switch, Diagram **C**. If the switch is pressed the lamp will turn off.

A common use of NOT gates is for emergency stops on engineering machines. If the switch is pressed whilst the machine is in use, the machine stops.

Using NOT gates in electronic circuits

The 40106 Hex inverter IC is a commonly used NOT gate. This is called Hex because it has six inverters on the IC. The pinouts are shown in Diagram **D**. Diagram **E** shows a simple circuit that includes this IC. The LED will be lit until the PTM switch is pressed. When the switch is released, the LED will illuminate again.

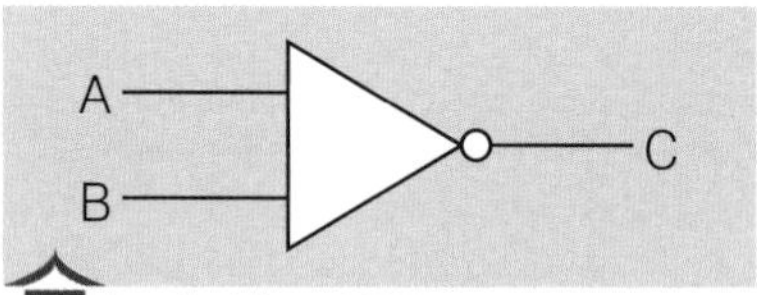

B *Symbol for a NOT gate*

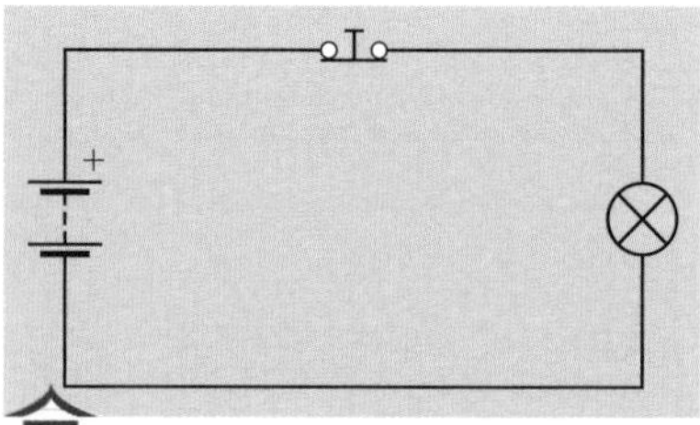

C *A mechanical NOT gate: if the switch is pressed, the lamp will turn off*

Connecting the outputs

In the examples given in topics 3.4, 3.5 and here, the logic gates have been connected directly to an LED to indicate that the gate is working.

Logic gate ICs will output the full voltage of a circuit but only a very low current. This means that the output from a logic gate IC cannot be used directly to drive high current devices such as lamps, motors or buzzers. In these cases a **transducer driver** such as a transistor (see topic 2.2) or thyristor (see topic 3.1) must be used, as shown, for example, in Diagram **F**.

Solving problems with logic gates

Logic gates are often used in combination in order to produce a desired outcome. For example, consider an alarm system that has a key-operated master switch and four pressure sensors placed around a building, Diagram **G**. The alarm must sound if any one OR other of the switches is tripped but only if this happens AND the key switch is in the on mode.

Objectives

- Describe the operation of a NOT gate.
- Explain why a transducer driver may be needed when using a logic gate IC.
- Explain how multiple logic gates may be needed to solve a problem.

A *Truth table for a NOT gate*

A	Q
0	1
1	0

Key terms

NOT gate: a logic gate that inverts the input to produce an output.

Inverter: a digital device that turns a high input into a low output and vice versa.

Transducer driver: a device such as a transistor or thyristor that can provide a high power output.

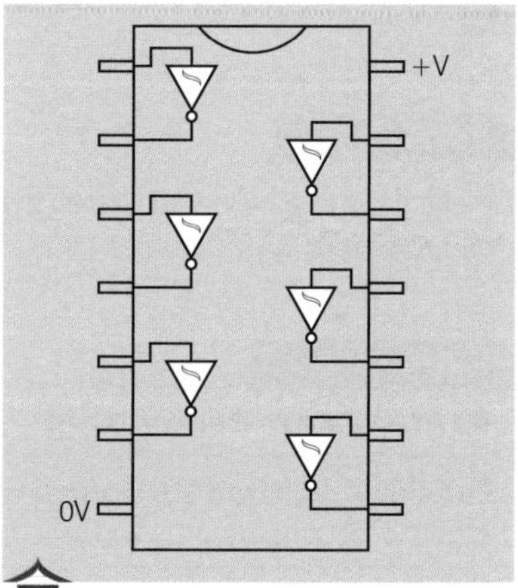

D *Pinouts for a 40106 Hex inverter*

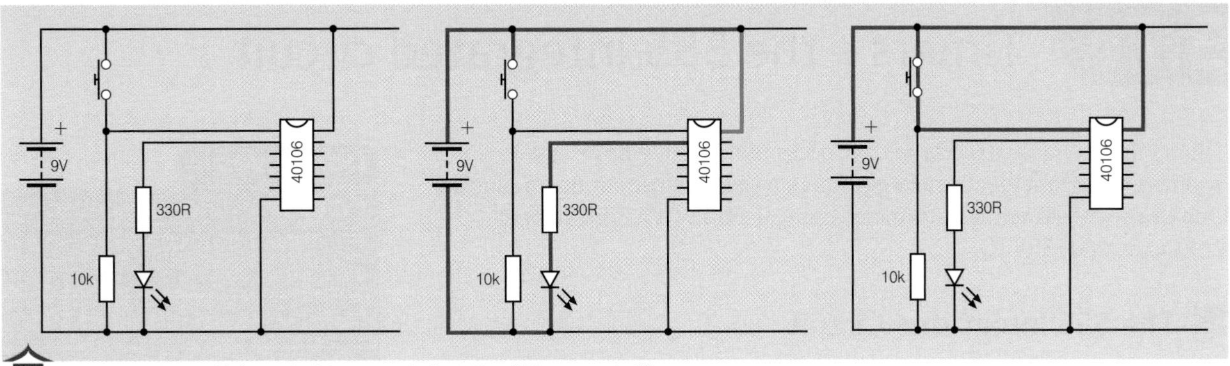

E *NOT gate circuit: if the switch is pressed, the LED will be turned off*

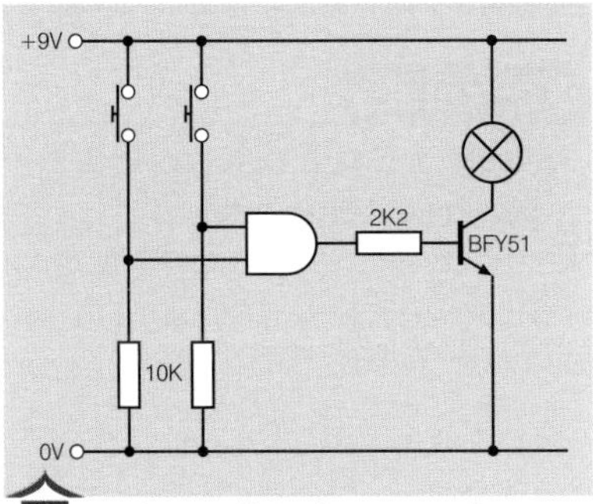

F *Logic gate linked to a transducer driver*

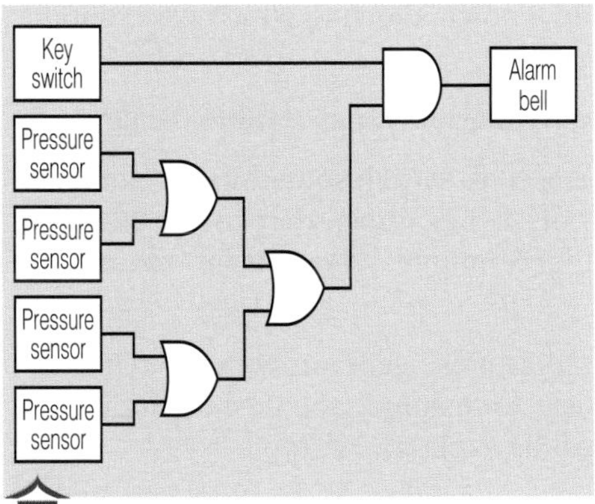

G *Logic diagram for an alarm system*

Activity

A tomato grower wishes to automate the watering of his plants. His watering system should fulfil the following conditions:

- The greenhouse must be light (the light sensor gives a 1 when in light conditions).
- The soil must be dry (the moisture sensor gives a 0 when in dry conditions).
- The temperature must be above 0 centigrade (the temperature sensor gives a 1 when above 0 centigrade).

In addition, an override switch must be fitted to allow watering at any time. Using the logic gate symbols, show how this could be achieved.

AQA Examiner's tip

When designing your circuit, consider whether different types of logic gate may allow the circuit to be simplified.

Summary

A NOT gate inverts the input signal.

The output from logic gate ICs can have the full voltage of the circuit but only low current. To operate high current devices they need to be used in combination with a transducer driver.

Multiple logic gates can be used in combination to solve problems.

kerboodle!

3.7 Timers – the 555 integrated circuit

Timers are a widely used type of process block. They have many applications, from electronic egg timers to snooze functions on alarms. One of the most common forms of timer used in GCSE Electronic Products is the 555 IC.

The 555 Integrated Circuit

The 555 IC is an eight-pin DIL package, Diagram **A**. It contains the equivalent of 40 transistors and resistors. Normally there is a dot marking the location of pin 1 and a notch showing which way round they should be placed.

Circuits use 555 ICs in two ways: monostable and **astable**, Diagram **B**.

In a monostable circuit, they are used to switch something on for a set amount of time and then turn it off. For example, they might be used to turn the siren of an alarm on for 5 minutes, then turn it off until it is reset.

In an astable circuit, they are used as pulse generators, to switch something on and off continuously, for example the flashing indicator light on a car. Astable circuits will be explained in topic 3.9.

Characteristics of the 555 IC

Diagram **C** shows the pinouts of a 555 IC. The threshold (pin 6) and discharge (pin 7) are the timing pins that are linked to the capacitor-resistor combination.

The 555 IC is a digital component normally powered using a 9 V supply or PP3 battery. As explained in topic 3.5, it has an upper threshold and a lower threshold to recognise whether an input is high or low. The upper threshold is normally two-thirds of the supply voltage (6 V). The lower threshold is normally one-third of the supply voltage (3 V).

The component normally uses 2 V internally. This means that the maximum output voltage at pin 3 would be 9 V – 2 V = 7 V. It is capable of operating with currents of 200 mA.

Outputs from the 555 IC

Diagram **D** shows how the output from a 555 IC can be directly linked into other ICs. However, the current output is not sufficient to drive high current devices such as lamps or motors. In these cases a transducer driver such as a transistor (see topic 2.2) must be used.

A *555 IC*

Objectives

- Explain the difference between the output of monostable and astable circuits.
- Identify the pinouts on a 555 IC.
- Describe the characteristics and use of a 555 IC.

Key terms

Astable: a circuit that provides a pulsing output signal.

Using IC sockets

ICs such as the 555, 741 op amp and the logic gates can be damaged by heat from a soldering iron. If they are soldered directly onto a printed circuit board (PCB) and do not work, they can be difficult and time-consuming to remove. For this reason, it is a good idea during GCSE project work to use IC sockets. These are available in sizes from 6 to 40 pins, and have a notch to indicate the position of pin 1. They fit in the same position on the PCB as the IC would.

When using IC sockets, the pins should never be bent over prior to soldering to hold them in place. If necessary, masking tape could be used to hold them in position.

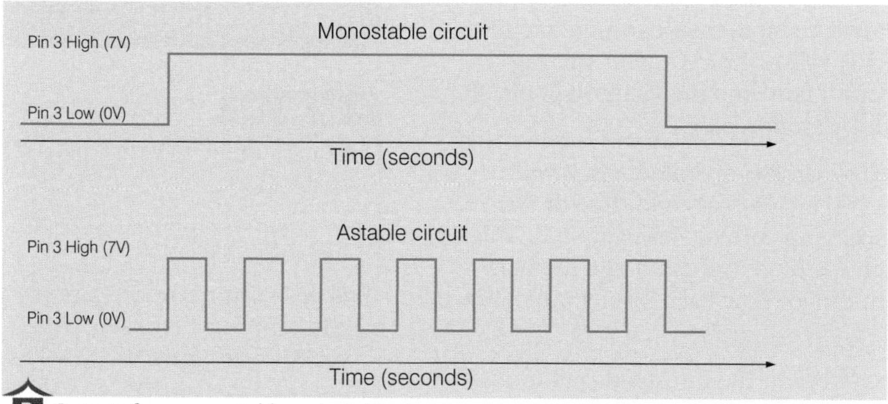

B *Outputs from monostable and astable circuits*

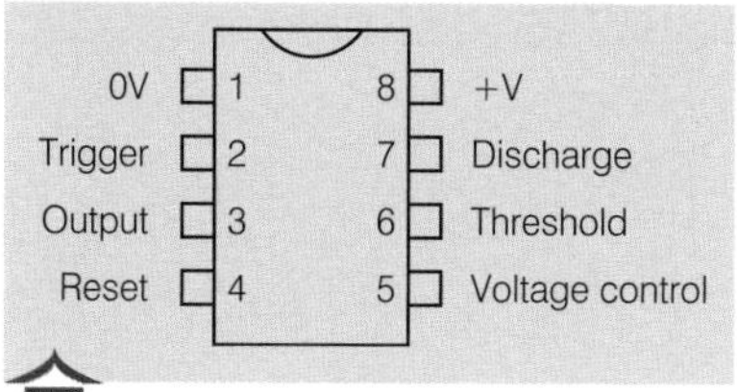

C *Pinouts for a 555 IC*

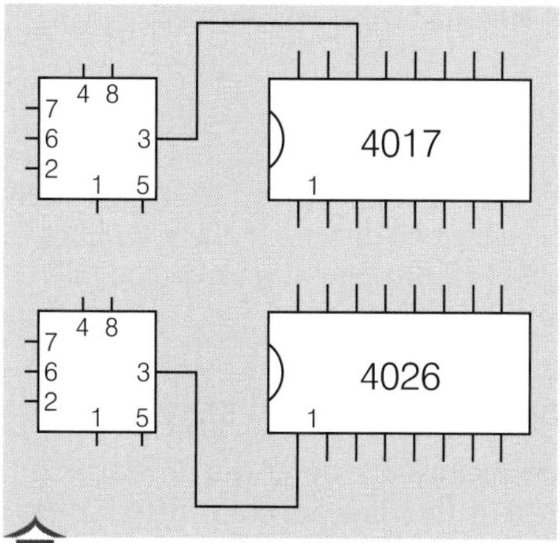

D *Outputs to a 4017 IC and a 4026 IC*

Summary

A 555 IC is a digital device that can be used as part of a monostable or an astable circuit.

A monostable circuit produces a high output for a set period of time. An astable circuit produces a continuous pulsed output.

With a 9 V supply a 555 IC can produce an output of 7 V. To operate high current devices it needs to be used in combination with a transducer driver.

Activity

Make lists of all the electronic devices you know that use timers and pulsed outputs.

AQA Examiner's tip

Memorise the pinouts for the 555 IC.

3.8 Timers – monostable circuits

How the 555 IC works in a monostable circuit

Diagram **A** shows the 555 IC in a monostable circuit. The schematic of the 555 IC on the circuit diagram is presented to make drawing easier and does not correspond to the pin locations of the real world component, but the pin numbers are shown. Note that pins 6 and 7 are connected directly to each other, which is a characteristic of a monostable 555 IC circuit.

When the circuit is first turned on, the voltage at the trigger (pin 2) will be high – this means that it will be more than two-thirds of the battery voltage. The output (pin 3) will be low, 0 V, and **sinking** current. This means that current will flow into the IC through pin 3. This allows the buzzer to sound, Diagram **A part ii**.

When the PTM switch connected to pin 2 is operated, pin 2 will become low – this means that it will be less than one-third of the battery voltage. The IC is now **sourcing** current through pin 3. This means that current flows through the IC, out of pin 3 and the LED lights, Diagram **A part iii**. Current cannot flow back through the buzzer due to the diode in the circuit.

Pin 3 will remain high until the voltage at the threshold (pin 6) goes high. This occurs when the time constant of the capacitor-resistor combination has been reached. When this happens, pin 3 will return low, to 0 V. The LED will turn off and the buzzer will turn on.

Pull up resistor

In topic 3.5 we looked at the use of a pull down resistor to ensure that an IC did not register a false high reading. Similarly, when using a 555 IC the trigger (pin 2) is normally connected directly to the 9 V rail or supply using a **pull up resistor**. This ensures that it receives the full voltage of the battery and registers as high.

Determining how long the circuit will stay on

The time period for which the 555 monostable timer will provide an output from pin 3 is determined by the **time constant**. This is governed by the capacitor-resistor combination used, Diagram **B**. This was explained in topic 1.9, and is calculated using the formula:

Time constant (seconds) = Capacitance (farads) × Resistance (ohms) = C × R

For example, for C = 470 μF and R = 100 KΩ:

Time = 0.000 47 × 100 000 = 47 seconds

Objectives

- Describe the sequence of operation of a monostable circuit using a 555 IC.
- Explain why a pull up resistor is used.
- Calculate the time constant of a monostable circuit.

Key terms

Sinking: current flows into pin 3 of the 555 IC.

Sourcing: current flows out of pin 3 of the 555 IC.

Pull up resistor: a resistor used to ensure an input receives the full voltage of a supply.

Time constant: the time to charge a capacitor used in series with a resistor.

The output time from the circuit in Diagram **A** has been made adjustable by using a variable resistor in series with the fixed value resistor. This arrangement is better than just using the variable resistor so that if the variable resistor is set to 0R, a short circuit does not occur that might set fire to the variable resistor. The minimum value of the fixed value (pull up) resistor used should be 1 KΩ. The maximum value of this resistor can be up to 1 MΩ. This gives a practical operating range of a monostable circuit of between about 0.1 seconds and 1000 seconds.

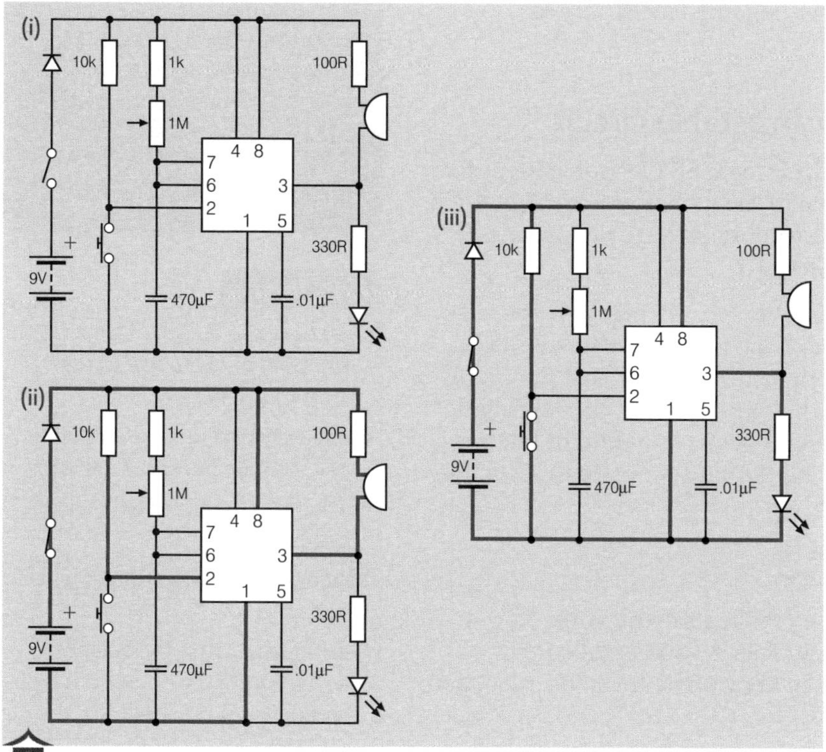

A *Monostable circuit in operation*

Remember

Sequence of operation: monostable circuit

Circuit switched on

Pin 2 high, pin 3 low, buzzer sounds

PTM switch pressed

Pin 2 low, pin 3 high, buzzer off, LED lights

Capacitor charges through R1 for the time constant

Pin 6 goes high

Capacitor discharges into pin 7

Pin 3 low, LED off, buzzer sounds.

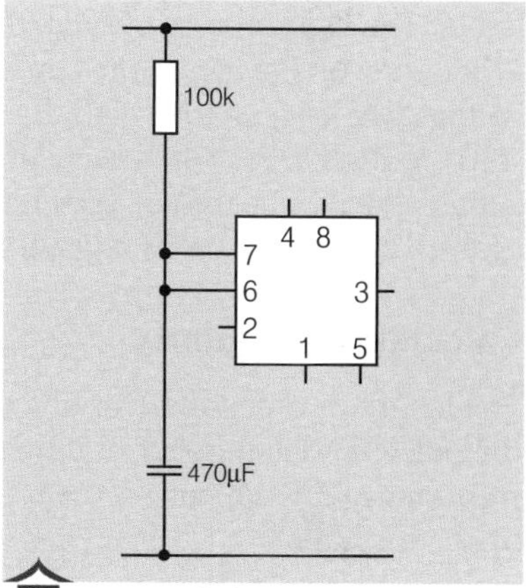

B *Circuit including a pull up resistor*

Activities

1. What value of resistors and capacitors would you use to create a suitable time constant for a 4-minute egg timer?

2. This activity can be carried out using either computer software that simulates the operation of circuits or a breadboard model of a circuit. Make a model of the circuit shown in Diagram **A**, substituting the resistor and capacitor that you calculated for the egg timer. Measure how long the timer gives the activity.

AQA Examiner's tip

You should memorise the sequence of operation of a monostable circuit.

You need to be able to recognise a monostable circuit. These often appear in exam questions.

You should be able to identify a pull up resistor and explain why it is needed.

Summary

The time period of the output for a monostable circuit including a 555 IC is determined by the time constant. This is determined by the capacitor and resistor used.

A pull up resistor is used so that an input on a digital device receives the full voltage of the supply, to ensure that it registers high.

3.9 Astable circuits

An astable circuit is a very useful process block. It is also called a pulse generator as it provides an output that pulses between high and low voltage, as shown in Diagram **A**.

Astable circuits have many applications. They are used for flashing lights, to produce sound from speakers, to control the speed of motors and to provide a clocked input to other building blocks such as counters.

How the 555 IC works in an astable circuit

The most common form of astable circuit in GCSE Electronic Products uses the 555 IC, Diagram **B**. Compared to the monostable configuration (topic 3.8), there is no switch to the trigger to activate the pulsing. It starts as soon as the supply is connected.

When the circuit is turned on, the output is initially high and the LED lights. The capacitor charges up through the timing resistors R_1 and R_2. When the voltage across the capacitor reaches two-thirds of the supply voltage, the trigger goes high, the output goes low and the capacitor starts to discharge. When the voltage falls below one-third of the supply voltage the output goes high and the capacitor begins to charge again. The process keeps repeating.

Calculating frequency

The length of time taken for one pulse is called the **duration** of the pulse, *T*. The number of pulses made per second is called the **frequency**, *f*. Frequency is measured in hertz (Hz). It can be calculated by the formula:

$$f = \frac{1.44}{(R_1 + 2R_2)} \times C$$

For example, in the circuit shown in Figure **B**, R_1 = 1 KΩ, R_2 = 39 KΩ and C = 10 mF. This gives:

$$f = \frac{1.44}{(R_1 + 2R_2)} \times C = \frac{1.44}{1000 + (2 \times 39\,000)} \times 0.000\,01 = 1.8 \text{ Hz}$$

The frequency is the number of times that the LED will flash each second. The frequency can be altered by changing the values of the capacitor (C) and the resistors (R_1 and R_2). The frequency will increase as the value of the capacitor increases or the total value of the resistors decreases.

In some applications, such as a metronome, the user might want to adjust the frequency of the circuit. This can be achieved by either substituting R_2 with a variable resistor or by adding a variable resistor in series with R_2. The second method will protect the IC from damage if the variable resistor is ever adjusted to 0 Ω.

Objectives

- Explain the function of an astable circuit.
- Describe the configuration of an astable circuit using a 555 IC.
- Adjust the frequency of the pulses in the 555 IC astable circuit.
- Calculate the mark/space ratio and explain what this means.

Key terms

Duration: the length of time from the start of one pulse to the start of the next.

Frequency: the number of pulses per second, measured in Hertz.

Mark time: the time that a pulse output is high.

Space time: the time that a pulse output is low.

Mark/space ratio: the balance between the time a pulse is high and the time it is low.

Remember

Sequence of operation: astable circuit

Circuit switched on

Pin 2 low, pin 3 high, LED lights

Capacitor charges through resistors R1 and R2

Pins 2 and 6 go high

Pin 3 low, LED off

Capacitor discharges through R2 into pin 7

Pin 2 low and the process repeats.

Mark/space ratio

The time for which the output is high (the **mark time**) and the time for which the output is low (the **space time**) are controlled by the timing resistors and capacitors in the circuit. The time high can be calculated using the formula time high = $0.693\ (R_1 = R_2)\ C$. The time low can be calculated by using the formula time low = $0.693\ R_2\ C$. The **mark/space ratio** is worked out by:

mark/space ratio = time high/time low

A circuit with a mark/space ratio of 1:1 would have equal on and off times. It is possible to create astables with unequal mark/space ratios by modifying the circuit.

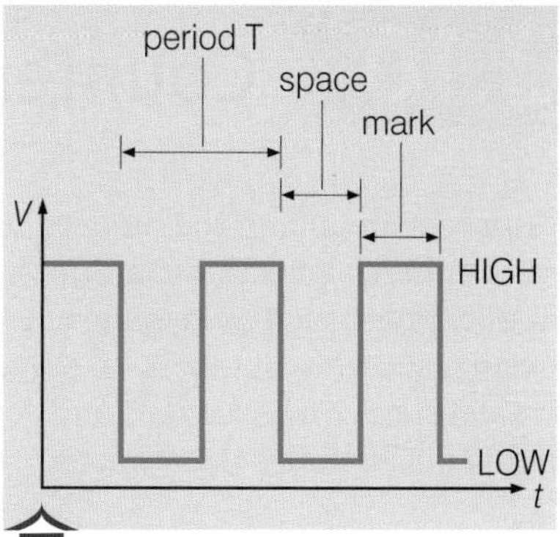

A *Astable circuit*

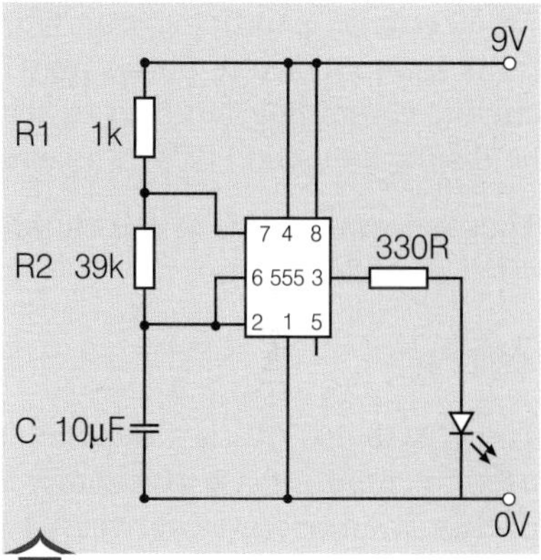

B *Astable 555 circuit*

Using a monostable 555 IC to control an astable 555 IC

A monostable circuit can be used to control an astable circuit, Diagram **C**. This provides a pulsed output for the set time period after the monostable is triggered.

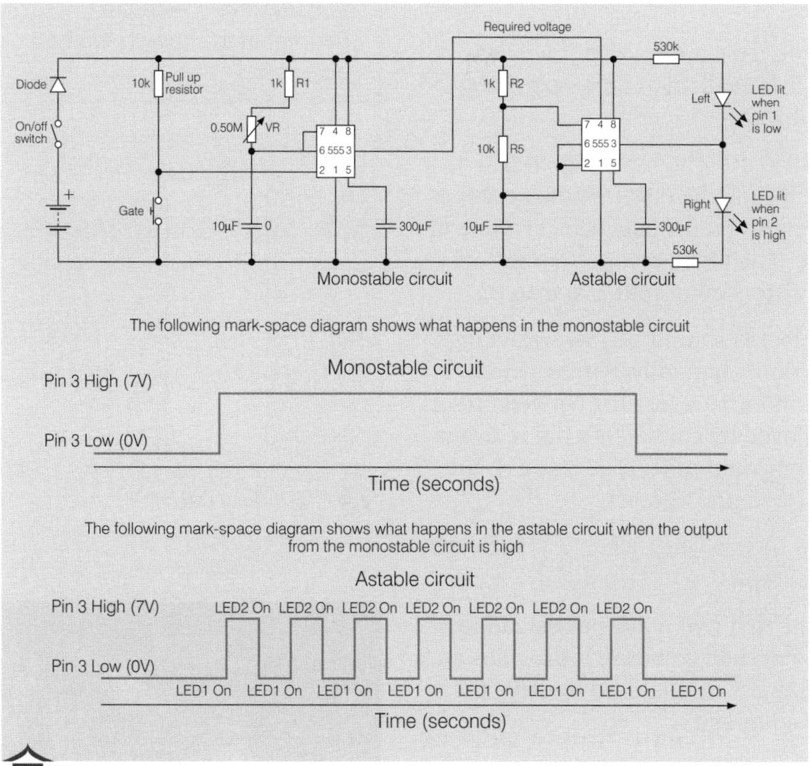

C *Monostable circuit controlling an astable circuit*

Activity

Calculate the frequency of an astable circuit If R_1 is 1 KΩ, R2 is 18 KΩ and C is 0.1 µF.

AQA *Examiner's tip*

Make sure you can recognise the astable/pulse generator circuit diagram. You need to be able to identify which components control the frequency.

Summary

An astable circuit generates pulses, and can be constructed using a 555 timer IC.

The frequency of the pulses depends upon the values of the timing resistors and capacitors.

The mark/space ratio is the ratio of the on time to the off time.

kerboodle!

3.10 Counters 1

Counters are one of the most commonly used process blocks in commercial electronic products. They have a wide range of applications, such as keeping count of the number of times an event occurs, showing numbers on seven-segment displays (used on digital watches, microwave ovens and so on), and flashing a sequence of lights in Christmas decorations.

Objectives

Explain how to use the 4017 IC as a decade counter and an event counter.

Key terms

Seven-segment display: visual display arranged to show decimal numbers.

CMOS: complimentary metal oxide semiconductor.

Decade counter: a counter that outputs the results in base 10.

Switch bounce: when mechanical switches make multiple (unintended) contacts whilst being used.

Electronic counting circuits

An electronic counting circuit works by counting the pulses it receives from an input. These pulses may come from pulse generators or input circuits containing switches or sensors. The result of the count can then be displayed using LEDs or a **seven-segment display**.

The key components of the counting circuits are digital ICs such as the 4017 or 4026.

4017 IC

The **CMOS** 4017 IC is a **decade counter**. This means that it can be used to count in base 10, from 0 to 9. It is available in a 16-pin DIL package and operates between 3 V and 15 V.

Diagram **A** shows the pinouts for the 4017 IC. The input signal is received in pin 14, the clock input. Each time the clock input registers a pulse, one of the ten outputs (0–9) comes on in sequence. Just one of the individual outputs is high at a time. The first clock pulse registers on output 0, the second clock pulse registers on output 1 and so on.

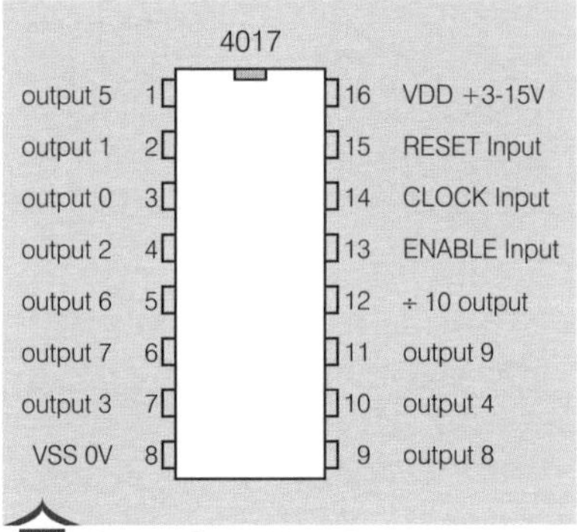

A *Pinouts for a 4017 IC*

For normal counting the reset input (pin 15) should be low; when high it resets the count to zero. This can be done manually, with a switch between reset and the positive supply and a 10 k resistor between reset and 0 V. Counting to less than 9 is achieved by connecting the relevant output to reset. For example, to limit the count to 0, 1, 2, 3 and 4, the output for 5 (pin 1) would be connected to pin 15 reset.

The clock enable input (pin 13) should be low since when it is high the clock pulses are ignored and the count stops.

The ÷ 10 output (pin 12) is high for the first five input pulses and low for the next five. This means that when decade counting it provides an output at one-tenth of the clock frequency. It can be used to drive the clock input of another 4017 (as a tens counter).

Using the 4017 IC with an astable input

Diagram **C** shows a counter circuit using the 4017 IC. The pulses at the clock input are being provided by an astable circuit. LEDs have been connected to the output pins. They will ripple on and off in sequence at a speed determined by the frequency of the clock pulse.

B *Christmas lights use counters to flash in sequence*

Using the 4017 IC as an event counter

The signal to the clock input can be provided by a mechanical switch or sensing circuit. This turns the circuit into an event counter, Diagram **D**. It should be noted that the mechanical switch has first of all been connected to trigger a monostable with a short time period (at pin 2). It is the output of the monostable (pin 3) that provides the clock input to the decade counter. This is to overcome the problem of **switch bounce** with mechanical triggers. This is explained in topic 3.11.

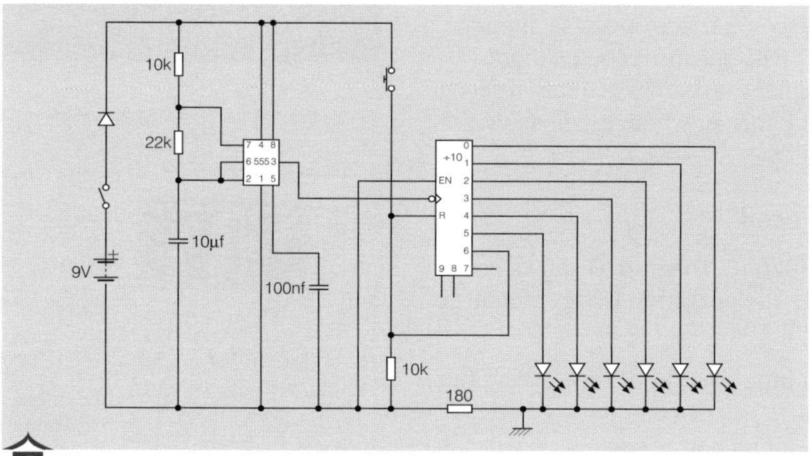

C *Using a counter circuit to light a series of LEDs*

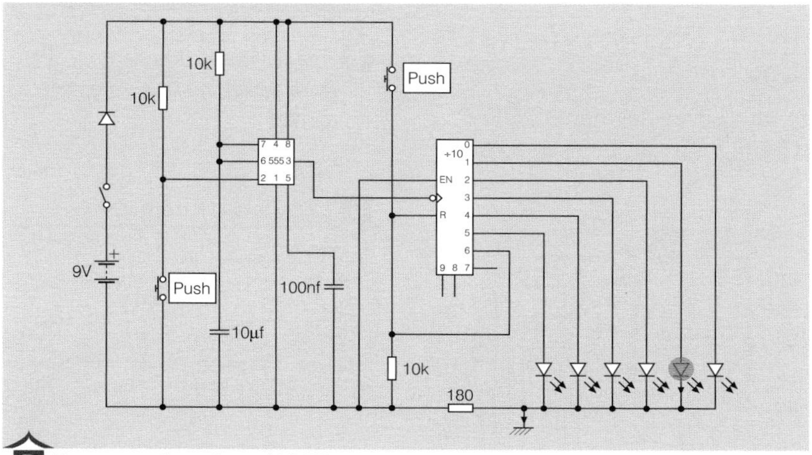

D *Event counting using a 4017 IC*

Activity

This activity can be carried out using either computer software that simulates the operation of circuits or a breadboard model of a circuit. Create the circuit shown in Diagram **D**. Work out how to limit the count to six, by using the reset input.

AQA Examiner's tip

You should be able to draw the missing connections on a circuit diagram of a 4017 IC to limit the count to six.

Summary

A 4017 IC has ten outputs. It produces a sequential ouput that can be cascaded or limited to fewer than ten outputs by linking an output to the reset.

3.11 Counters 2

4026 decade counter and seven-segment display driver

A disadvantage of the 4017 IC is that if a number needs to be displayed, the outputs will need to be converted into the display format. The CMOS 4026 IC is a decade counter and seven-segment display driver, packaged in a 16-pin IC. This means it counts input clock pulses and provides an output signal that can be fed into a seven-segment display such as that shown in Photo **A**, where it will show the number in decimal form.

How the 4026 decade counter is used

Diagram **C** shows the pinouts for the 4026 IC. Diagram **D** shows how this is used in a decade counter circuit. The operation is very similar to that for the 4017 IC explained in topic 3.10:

- For counting 0–9 the reset (pin 15) input should be low. Taking the reset high resets the count to zero.
- The disable clock input (pin 2) should be low. If taken high it disables counting so that the clock pulses are ignored and the count stops.
- The ÷ 10 output (pin 5) can be used to drive the clock input of a second 4026 to allow counting up to 99.

To provide an output that can be shown on a seven-segment display, the enable display (pin 3) input should be high. It can be used to give a blank display if taken low.

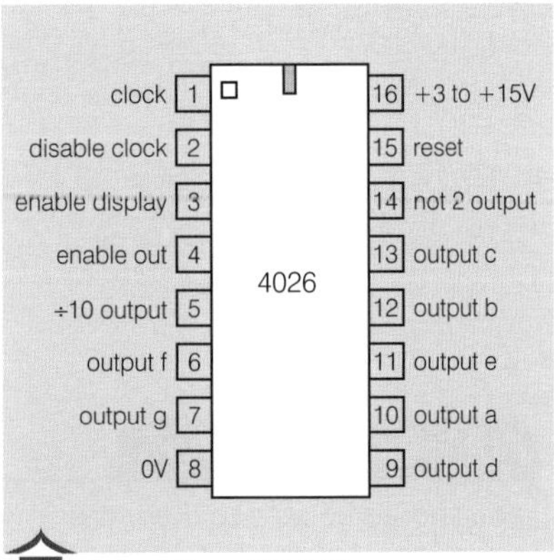

C *Pinouts for a 4026 IC*

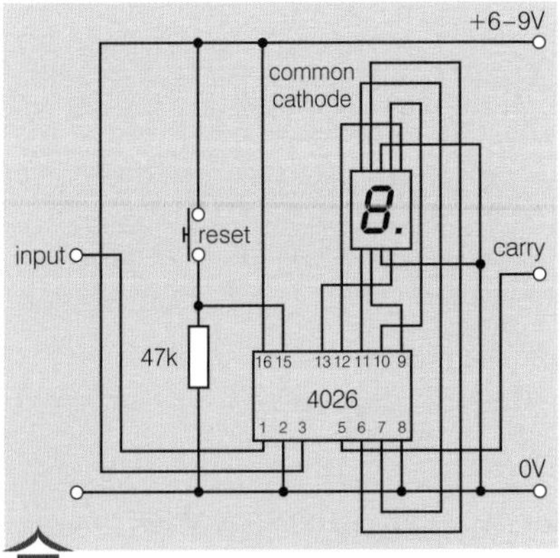

D *4026 IC set up to count 0–9*

Objectives

Explain how to use the 4026 IC as a counter.

Explain the need to use a debounce system.

A *The numbers on this alarm clock are each shown on a seven-segment display*

B *Binary to decimal conversions*

Decimal	Binary			
	Eights	Fours	Twos	Units
0	0	0	0	0
1	0	0	0	1
2	0	0	1	0
3	0	0	1	1
4	0	1	0	0
5	0	1	0	1

Switch bounce

When a mechanical switch is pressed, the switch's contacts may bounce against each other, turning the switch on and off in rapid succession, Diagram **E**. If connected directly to the clock input of a counting device, switch bounce is misinterpreted as multiple input pulses (pressing the switch several times). This results in the count advancing higher than intended. A debounce circuit is needed to clean up noisy input pulses from mechanical switches. Topic 3.10 showed a 555 monostable timer set on a very short time period to overcome the problem.

Alternatively a Schmitt trigger circuit can be used, such as the 40106 IC which is available as a 14-pin IC containing six separate Schmitt NOT gates.

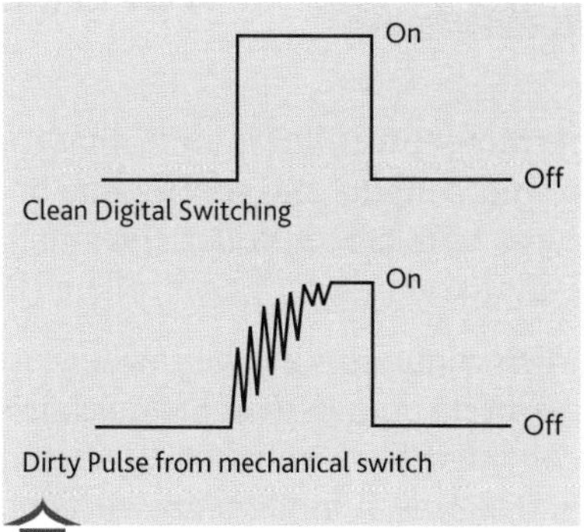

E *Effect of bouncing on a mechanical switch*

Binary counting

There are other electronic circuits which can be used as counters. As with all electronic circuits, these count in **binary**. This is the simplest possible counting system because it uses just two digits, 0 and 1. In a binary number each digit represents a multiple of two (1, 2, 4, 8 etc), in the same way that each digit in decimal represents a multiple of ten (1, 10, 100, 1000 etc) Table **B**. For example, 1001 in binary is equal to 9 in decimal. Each binary digit is called a **bit**, so 1001 is a 4-bit number.

An additional circuit, called a **decoder**, is needed to convert the numbers from binary form to display them as decimal numbers which people can understand. One of the advantages of using the 4017 IC or 4026 IC is that the decoding is carried out within the IC, making the circuit much simpler.

Key terms

Binary: number system used in digital devices with only two possible values for each digit, 0 and 1.

Decoder: device that converts binary into decimal format.

Activity

This activity can be carried out using either computer software that simulates the operation of circuits or a breadboard model of a circuit. Create the circuit shown in Diagram **D**. Use this to display each of the numbers 0 to 9 in sequence.

AQA Examiner's tip

You need to know which pin on the 4026 IC is the clock input and how to reset counters to 0.

You should be able to explain how to debounce the signal from a mechanical switch.

Summary

A 4026 IC can be used with an astable pulse generator to count from 0 to 10 using a seven-segment display.

Mechanical switches can be debounced by using a monostable timer set on a short time period or a Schmitt trigger.

kerboodle!

3.12 Microcontrollers 1

A **microcontroller** is an IC that can be programmed to respond to one or more inputs and control one or more outputs. The most common types of microcontroller encountered in GCSE Electronic Products are Peripheral Interface Controllers (**PIC**s) and picaxe ICs.

Microcontrollers are very widely used. For example, in the home they might be used in washing machines, remote controls and DVD players. In commercial environments, they can be used to control vending machines and alarms. A modern car can contain around 40 microcontrollers.

Objectives

- Explain what a microcontroller is.
- State what PIC stands for.
- Describe a range of applications where microcontrollers are used.
- Provide an appropriate, controlled voltage to operate a microcontroller.

Key terms

Microcontroller: a type of programmable microprocessor.

PIC: peripheral interface controller; or programmable interface controller; programmable integrated circuit.

Why use a microcontroller?

Microcontrollers are generally much more expensive than other ICs. However, a single microcontroller can be used to carry out the functions of several process blocks. For example, the process blocks for an alarm circuit might include:

- a comparator, to compare the signal from a light sensor to a reference value
- an AND gate, so that the alarm will activate only when the master on/off switch is on AND a sensor is activated
- an OR gate, so that the alarm will activate when any one of three sensors is activated
- a monostable circuit, so that the siren will sound for a set time of 5 minutes
- an astable circuit, so that the output from the siren will have two tones.

All of these functions could be carried out by one microcontroller IC, rather than five ICs (741, 4081, 4071 and two 555s). This means that the resulting circuit can be much simpler and smaller.

Common types of microcontroller

The microcontrollers used in GCSE Electronic Products are normally FLASH reprogrammable – this is shown by an F within the component number, for example 12F629 or 16F628. They can be reprogrammed as many as 100 000 times.

Flash microcontrollers are widely available in DIL packages with a range of sizes, including 8, 14, 18, 20, 28 and 40 pins. As the number of pins increases, both the cost and the numbers of possible inputs and outputs increase.

Some commercially used microcontrollers are one-time programmable (OTP) devices. They can be programmed once and if the program is wrong, the PIC cannot be used again. There are industrial versions of OTP devices that allow an existing program to be erased by special types of ultraviolet light, so that they can be reprogrammed. However, there would be safety risks from using this type

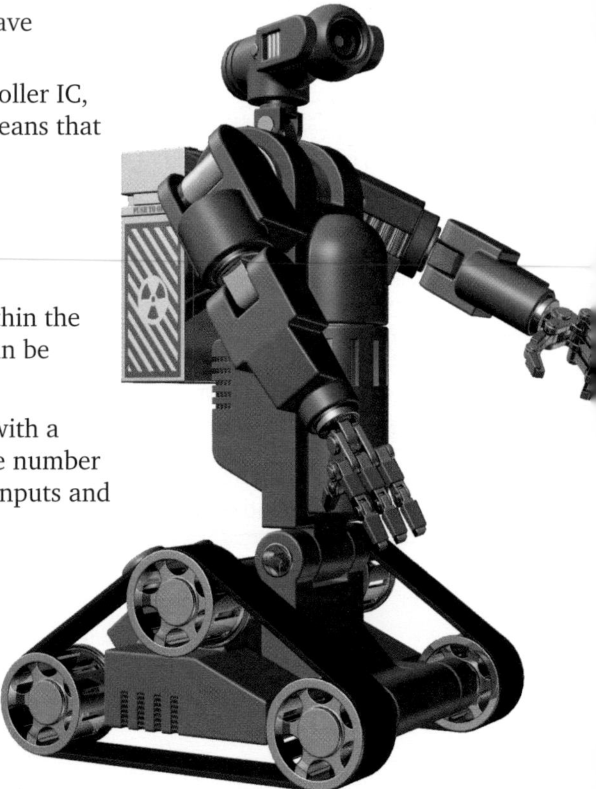

A *Robots are often operated using microcontrollers*

of light in a normal workshop environment. These ICs have a C within the component number. The advantage of OTP devices is that they can be less prone to the program being damaged than the FLASH devices.

Power supply

PIC and picaxe microcontrollers are very sensitive to voltage. They only operate with a voltage of 3.0 to 5.5 V DC. This is normally provided by three AA cells (3 × 1.5 V = 4.5 V).

If the power source is attached in reverse, this will damage the microcontroller. To avoid this, a diode can be used in series with the power supply to provide polarity protection. As the diode has a voltage drop of 0.7 V across it, four AA cells can be used. This gives a voltage of (4 × 1.5 V) – 0.7 V = 5.3 V.

A sudden voltage drop can scramble the program used to control a microcontroller. To avoid this it is recommended practice to use decoupling and smoothing capacitors. In addition, a reliable, accurate supply of exactly 5 V can be obtained by using a 7805 voltage regulator.

Activity

Using the internet, find examples of three applications where microcontrollers are being used to provide process control.

Summary

Microcontrollers are programmable microprocessors.

Common types of microcontroller include peripheral interface controllers (PICs) and picaxe ICs.

A single microcontroller can be programmed to carry out several process blocks in a circuit, reducing the number of ICs needed.

PIC and picaxe microcontrollers need a voltage of 3.0 to 5.5 V.

Remember

PIC

PIC is the name for a range of products developed by Arizona Microchip Technology. It stands for peripheral interface controller. Many people use the alternative names programmable interface controller or programmable integrated circuit.

Remember

Processing speed

Most modern microcontrollers contain an internal resonator that operates at 4 MHz. This means that they work at 4 million cycles per second!

AQA Examiner's tip

You should be able to describe the advantages and disadvantages of using microcontrollers.

kerboodle!

3.13 Microcontrollers 2

Inputs and outputs

A microcontroller always has at least one input and one output and depending upon the number of pins, it may have several of each type, Diagram **A**. For some microcontroller ICs, some pins can be used as either inputs or outputs, depending on what they are told to be by the program. A single microcontroller can handle several different inputs and outputs at the same time.

The outputs are digital. This means that the signal can be on or off, like the output from a simple switch. However, the inputs to a PIC may be either digital or analogue.

Analogue signals can vary continuously in size. For example, the signal from a temperature sensor may increase or decrease on a scale. Many PIC and picaxe ICs have analogue-to-digital conversion (ADC) built in to one or more pins, so that analogue sensors such as LDRs and thermistors can be linked directly to them.

The output from a microcontroller can be directly linked into other ICs. However, the current output is not sufficient to drive high-current devices, such as lamps or motors. In these cases a transducer driver such as a transistor (see topic 2.2) or thyristor (see topic 3.1) must be used.

Objectives

Identify the inputs and outputs on a microcontroller.

Describe the different approaches used to program microcontrollers.

Key terms

Program: the series of instructions that tell a microcontroller what to do.

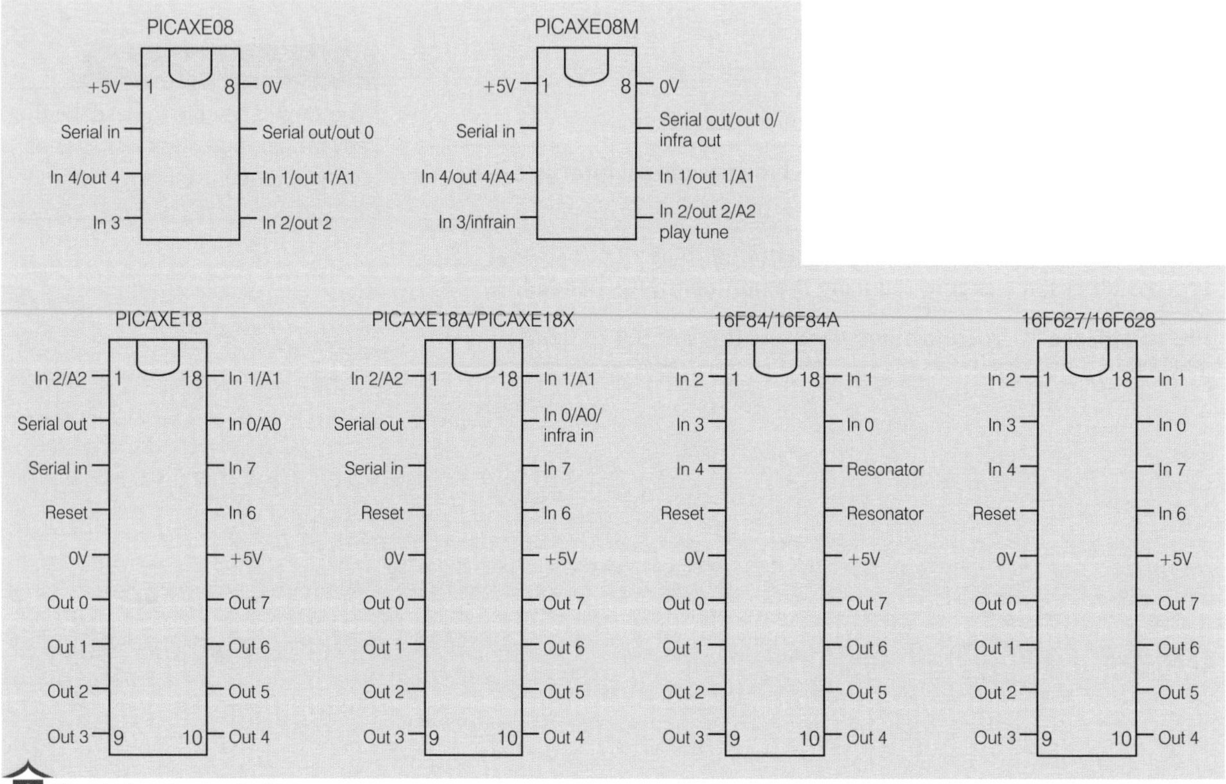

A *Pinouts for microcontrollers*

Programming a PIC

Microcontrollers vary in the amount of memory that they have available to hold the **program** that controls them. They read this program in a form called machine code. However, this is very difficult to understand, so programs are normally written in other languages which are converted using computer software to the machine code.

In industry, programs for microcontrollers are written using assembler code. This is complicated and hard to understand, but produces fast and efficient programs.

In schools, programs are written using either the BASIC programming language (Beginners' All-purpose Symbolic Instruction Code) or created as flowcharts (see topic 3.14). Both approaches allow subroutines for different process blocks to be included in a single program and can use either decimal numbers or binary numbers (see topic 3.11). The software used to create the flowcharts can automatically convert them into BASIC when preparing to program the microcontroller. These approaches need more memory to run than assembler code and are slightly slower, but for most practical applications in GCSE Electronic Products this is not noticeable.

To program a PIC microcontroller, it is placed in a piece of hardware called a programmer and the software program is downloaded from a computer. Picaxe microcontrollers can be programmed through a download socket that is included as part of the circuit design. This has the advantage that the IC does not have to be removed from the circuit for programming, reducing the risk of damage from handling or electrostatic charge from fingers, for example.

B *Example of a basic microcontroller program for a steady hand game*

green:	high 3	switch on the green LED
nolives:	if pin0=1 then amber	if the wire is touched goto amber
	goto nolives	go back to the beginning of nolives
amber:	high 2	switch on the amber LED
	wait 1	wait for one second to prevent a 'double hit'
firstlife:	if pin0=1 then red:	if the wire is touched goto red
	goto firstlife	go back to the beginning of firstlife
red:	high 1	switch on the red LED
	wait 1	wait for one second to prevent a 'double hit'
secondlife:	if pin0=1 then gameover	if the wire is touched go to gameover
	goto secondlife	go back to the beginning of secondlife
gameover:	high 0	switch on the buzzer
	wait 5	wait for 5 seconds
	low 0	switch off the buzzer
	low 1	switch off the red LED
	low 2	switch off the amber LED
	goto nolives	go back to the beginning of nolives

AQA *Examiner's tip*

You should be able use binary numbers in your control programs to control up to eight outputs.

Activity

Using the programming software available in your school, create a program that could be used to control a simple washing machine through a simple cycle. The inputs would be the door lock, the on switch and the temperature. The outputs would be linked to the water supply, the motor and a water heater. If the door is locked and the on button pressed, the water supply should be turned on for 5 minutes. If the water is less than 40° C the water heater should be turned on. Finally, the motor should then be turned on for 20 minutes to spin the clothes.

Summary

Microcontrollers have at least one input and one output, and may have several depending upon the number of pins they have.

Many common microcontrollers have ADC built into one of the input pins; the other inputs and outputs are normally digital.

The operating programs may be written in machine code, BASIC or using flowcharts.

kerboodle!

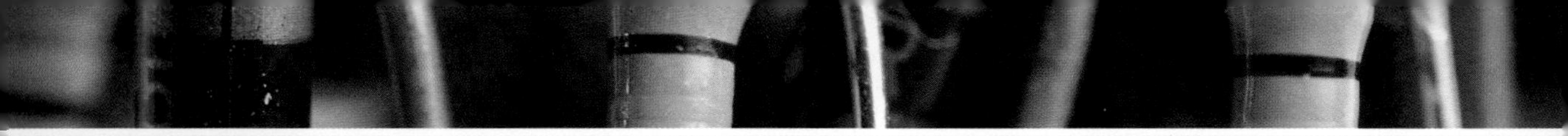

3.14 Flowcharts

Flowcharts

A **flowchart** shows the order in which a series of events are carried out. They are often used to write the programs that are carried out on microcontrollers, such as PIC and picaxe ICs.

A common mistake when designing a new product is to confuse flowcharts with systems diagrams. They show different things. A systems diagram shows the physical parts of a system – the equipment used to carry out the tasks. A flowchart shows the instructions that control what the system does. In this type of application, the flowchart could be thought of as providing the intelligence or program for the process box in a systems diagram.

Drawing flowcharts

Different symbols are used to represent the different possible types of event. Some of these are shown in Diagram **A**:

- A terminator is used for the start and end of the series of events.
- The process box is used when an instruction or action must be carried out.
- An input/output is used where information or items are added or given out.
- A decision represents a question or choice. It must have a yes or no answer.

The complete set of symbols and their meanings are listed in British Standard BS4058. The event or question that a symbol refers to is written inside the symbol.

The symbols are linked together by arrows which show the correct sequence of events. The arrows should always be either horizontal or vertical, never slanted.

When drawing flowcharts, the blocks and symbols should be kept a uniform size. If they are available, stencils can help in doing this.

Uses of flowcharts

Flowcharts can be used to create the programs for computer-controlled systems, such as microcontrollers, robots and computer-operated machines. For example, Diagram **B** shows the flowchart to control a simple, low-cost robot buggy. In the systems diagram it was determined that the buggy would have two contact switches and two motors.
The first step in developing the control program was to write out the instructions. When turned on, the robot buggy moves forward. When it touches a wall with the right sensor it stops, reverses a short way, then turns left and moves forward again until it touches another wall. When it touches a wall with the left sensor it stops, reverses a short way, then turns right and moves forward again until it touches another wall. These instructions were used to develop the flowchart.

Objectives

- Explain the difference between a flowchart and a systems diagram.
- Describe how to draw a flowchart.
- Describe how flowcharts are used.

Key terms

Flowchart: a diagram showing a sequence of operations or activities.

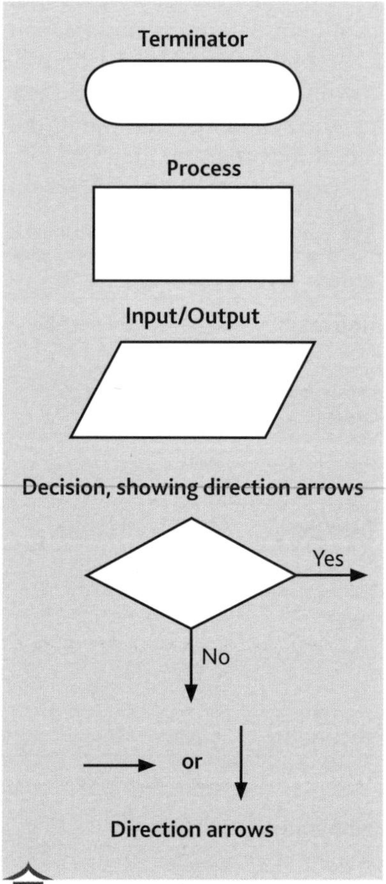

A *Some of the symbols used in a flowchart*

An important feature of flowcharts to control PIC microcontrollers are the wait commands. A PIC microcontroller can carry out 4 million operations per second. The wait commands are used to tell the microcontroller how long each output (or movement) should last. If these were not included, the microcontroller would only carry out that action until the next instruction was given – this could be as little as one four-millionth of a second.

Once the flowchart has been modelled and tested by the computer software, it is converted into a basic program and used to program the microcontroller.

Flowcharts can also be used to plan the sequence of actions during the manufacture of a product. This can be carried out as part of the development of the production plan, which will be covered in topic 8.12.

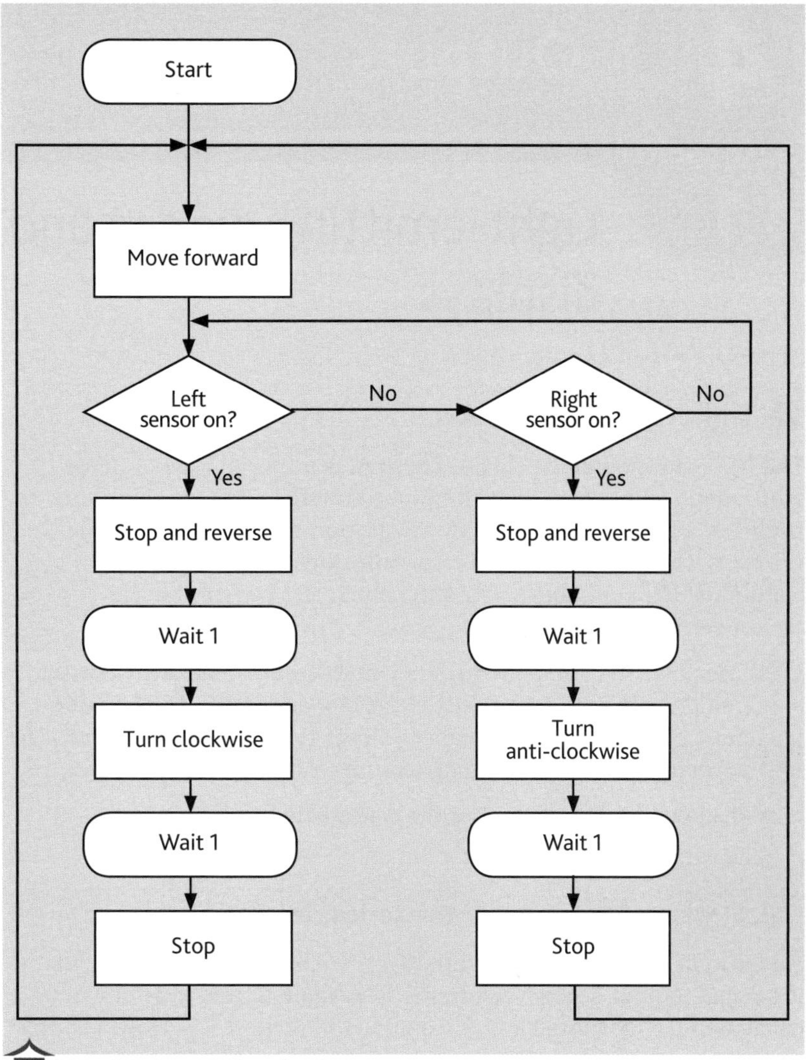

B *Flow chart to control a robot buggy*

Activities

1. Produce a flowchart to control the sequence of operations at a traffic light.
2. Produce a flowchart to control the automatic doors at the entrance to a supermarket. The doors have a sensor that will detect when someone approaches them and should open automatically. They should close automatically when the person has passed through them.

Summary

A flowchart shows a sequence of operations or events.

Each event in the sequence is shown inside a symbol. Each symbol has a specific meaning. The symbols are joined up by arrows, which show the direction of the sequence of tasks.

Flowcharts can be used to program computer-controlled systems and microcontrollers. They can also provide instructions for manufacturing tasks such as quality control, and to sequence tasks such as the making activities for an engineered product.

AQA Examiner's tip

Use a flowchart to model the operation of a PIC microcontroller in a circuit.

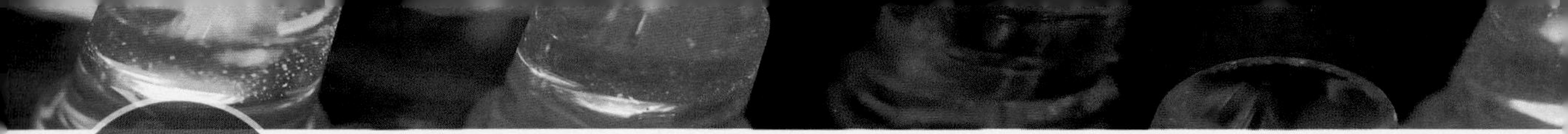

4 Outputs

4.1 Light-emitting diodes and seven-segment displays

Light-emitting diodes

The light-emitting diode (LED), Photo **A**, is a special type of diode which emits light when current flows through it. This makes it very useful as a visual indicator. They are used in all sorts of electronic products, for example as power on indicators. Compared to traditional filament lamps, LEDs use much less power and they do not get hot in use.

LEDs are polarised components. They must be connected the correct way round in a circuit to conduct electricity and light up. The **anode** lead must be connected to a higher voltage than the **cathode** lead. The LED normally has two identifying features to indicate its polarity:

- The anode leg is longer than the cathode leg.
- The cathode is indicated by a flat on one side of the LED case.

Calculation of the value of the series resistor

An LED typically requires around 20 mA of current. A series resistor is needed to limit excessive currents. The value of this resistor can be calculated using Ohm's Law, as explained in section 1.7. However, this must take into account that the LED needs about 2 V across it to work. For the circuit in Diagram **B**, the value of the current limiting resistor can therefore be calculated as follows:

$$\text{Resistance} = \frac{\text{Supply voltage} - 2\text{ V}}{\text{Current through LED}} = \frac{9\text{ V} - 2\text{ V}}{20\text{ mA}} = \frac{7}{0.02} = 350\ \Omega$$

Normal practice would be to choose the nearest preferred value of fixed resistor which is higher than the calculated need, such as a 360 Ω resistor or a 470 Ω resistor if a 360 Ω is not available.

Types of LED

LEDs come in a wide range of sizes, shapes and colours. The different colours are created by the different materials used to make them. Some LEDs can produce two or three colours.

A bi-colour LED has two LEDs combined in one package with just two leads, one connected forwards and the other backwards. Only one of the LEDs can be lit at one time.

The most popular type of tri-colour LED has red and green combined in one package with three leads. They are called tri-colour because when both LEDs are on, the mixed red and green light appears to be yellow.

Objectives

- Describe how an LED is used and positioned with the correct polarity.
- Calculate a suitable current limiting resistor to protect an LED.
- Describe the set-up of a seven-segment LED display.

Key terms

Anode: the positive leg of an LED

Cathode: the negative leg of an LED.

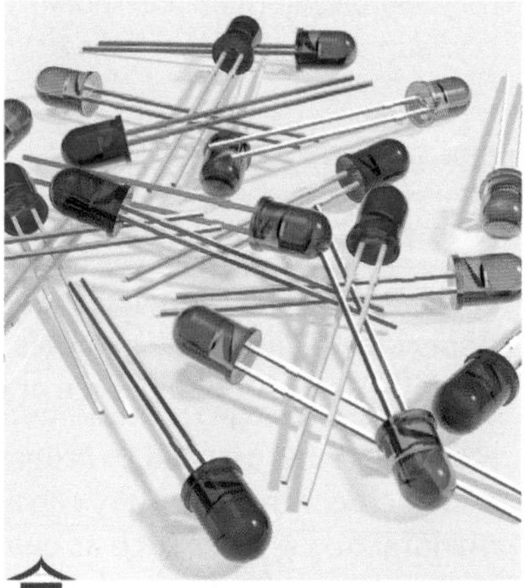

A *LEDs*

The middle leg is the common cathode for both LEDs and the outer legs are the anodes. These allow each LED to be lit separately or together.

LEDs are available which emit infra-red light, not visible to the human eye. This can be registered by light-dependent resistors (LDRs) of the same frequency. This pair of components is often used to create remote control systems.

Seven-segment display

A seven-segment display consists of seven small LEDs, a, b, c, d, e, f and g, arranged as shown in Diagram **C**. By lighting the segments in various combinations the numbers 0–9 can be displayed.

The display in Diagram **C** is a common-cathode type. This means that all the LEDs have the same cathode connection, but each anode is connected to a different pin. The segments light with high inputs to the anode pins and 0 V to the cathode connection. For example, to display the number 1, segments b and c must light, while all other segments must stay off. Therefore anodes b and c must receive high inputs whilst anodes a, d, e, f and g must receive low inputs.

Seven-segment displays are often used with the 4026 decade counter and seven-segment display driver, as shown in Diagram **C** in section 3.11. They can also be used with a PIC microcontroller that has seven available outputs.

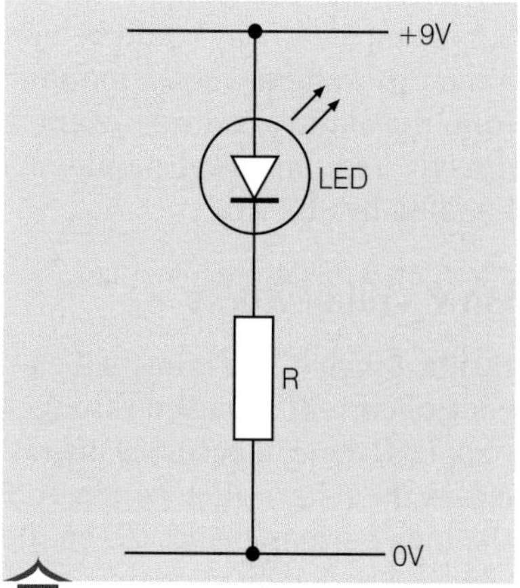

B *Using a current limiting resistor with an LED*

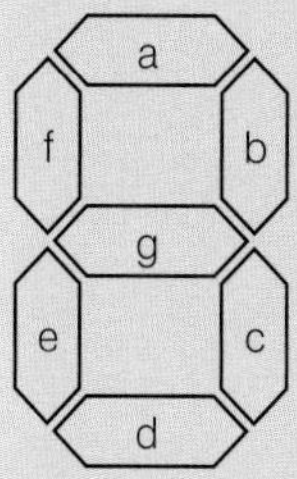

Segments							Decimal number shown
a	b	c	d	e	f	g	
1	1	1	1	1	1	0	0
0	1	1	0	0	0	0	1
1	1	0	1	1	0	1	2
1	1	1	1	0	0	1	3
0	1	1	0	0	1	1	4

C *Seven-segment display*

Activity

Draw out the numbers 0–9 as they would appear on a seven-segment display. Complete the table included in Diagram C to show which segments would need to be high and which segments would need to be low for all the numbers up to 9.

AQA Examiner's tip

Be sure that you can recognise the anode and cathode sides of LEDs, both as circuit symbols and on actual components.

Make sure you can calculate the value of the series resistor used to protect the LED.

Summary

An LED is a special type of diode that emits light when current flows through it in the correct direction.

LEDs must be protected by a series resistor.

A seven-segment display is an arrangement of LEDs used to show decimal numbers.

kerboodle!

4.2 Relays and opto-isolators

Relays

A **relay** is an output device. It is used to link electrical circuits without any electrical connection. Relays are used in a wide variety of products. For example, a relay could be used to activate the high-current sirens in a security system when a motion sensor is triggered. The output signal from the alarm's control system may be too small to drive the sirens directly and the power requirement of the sirens may be too high to be handled by an FET.

How a relay works

Inside a relay there is a small electromagnet or coil, Diagram **A**. A small control current is passed through the coil to energise it. When this happens, the coil becomes magnetic and attracts the armature, which closes the relay switch contacts. This acts as the switch to turn on the high current circuit. When the control current is switched off, the contacts spring open again.

Relays come in many shapes and sizes. They are usually PCB mounted and have either SPDT or DPDT switch types. The relay's switch connections are usually labelled COM, NC and NO.

- COM stands for Common (armature); it is the moving part of the switch.
- NC stands for Normally Closed; COM is connected to this when the relay coil is off.
- NO stands for Normally Open; COM is connected to this when the relay coil is on.

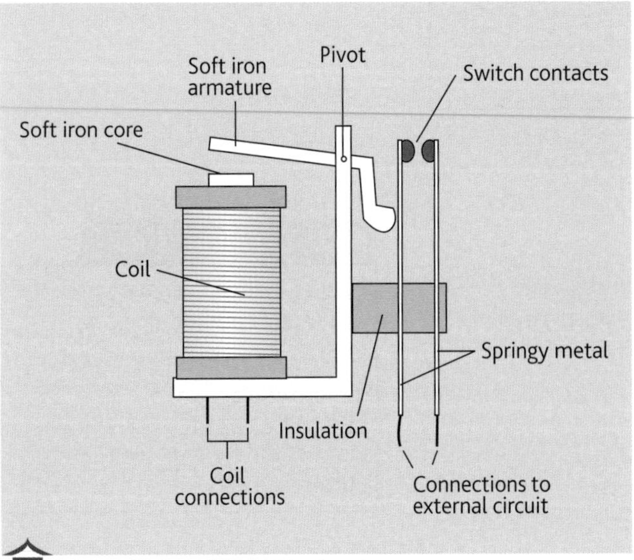

A *Schematic view of the inside of a relay*

Objectives

Explain how a relay is used as an interface device.

Explain the function of an opto-isolator and how it is used.

Key terms

Relay: an electromagnetic interface device.

Opto-isolator: a light-based interface device.

Remember

There is no electrical connection inside the relay between the two circuits. The link is magnetic and mechanical.

Using a relay as an output device

The most common uses of relays are to interface between a low-current control circuit and a circuit which requires a much greater power output than the control circuit can provide. For example, Diagram **B** shows a circuit diagram for some security lights that come on when it becomes dark. The relay forms the interface between the low-current LDR-based sensor circuit and the high-current power lamp circuit.

A disadvantage of relays is that their coils produce brief high voltage spikes when they are switched off. These spikes, called back emf, can destroy the transistors and ICs in a circuit. Back emf is due to the magnetic field collapsing around the coil of the relay when the current is stopped. To protect against this, a clamping diode is connected in reverse across the relay coil.

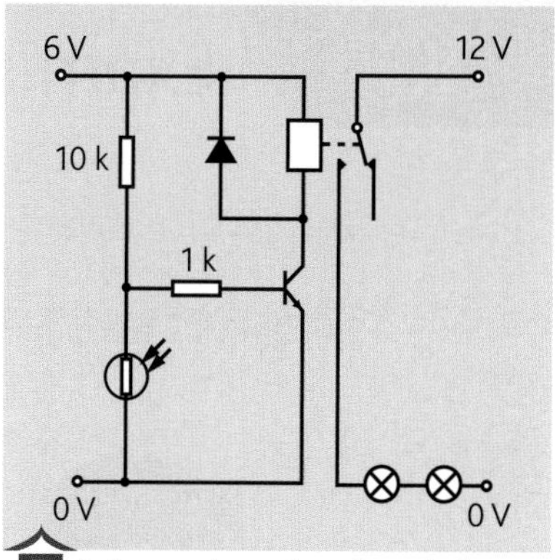

B *Relay used as an interface device*

Using a relay as a latch

A relay can also be used as part of a process block to latch a switching circuit. This uses a DPDT relay, as shown in Diagram **C**. The first set of SPDT contacts is being used to switch the secondary circuit. The second set of SPDT contacts is being used to create the latch by connecting the collector and emitter leads of the transistor together. This has created an alternative path for current to energise the relay coil in the event that the transistor is switched off.

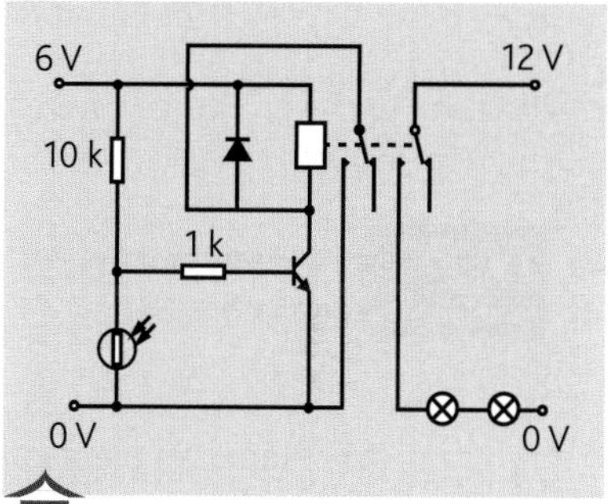

C *Relay used as a latch*

Opto-isolators

An **opto-isolator**, Diagram **D**, is another type of interface device which can be used to link circuits. However, this time the connection between the two circuits is infra-red (IR) light.

An opto-isolator consists of an IR-emitting LED, combined with a phototransistor in the same package. The two components are separated so that infra-red can travel across a gap but electric current cannot. Received IR at the phototransistor generates an output signal used to control a secondary circuit.

Opto-isolators are used in applications such as detecting the level of liquid in a tube or counting the number of rotations of a shaft.

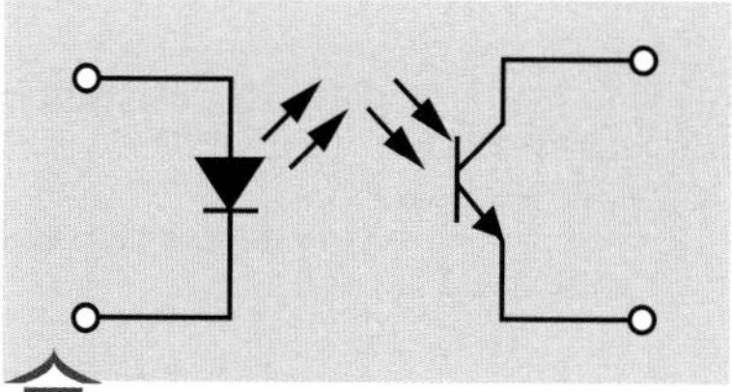
D *Opto-isolator*

Activity

This activity can be carried out using either computer software that simulates the operation of circuits or a breadboard model of a circuit. Create the circuit shown in Diagram **D**. Identify and test two ways of deactivating the latch.

AQA Examiner's tip

You need to use a clamping diode with a relay to protect other components from back emf.

You should be able to complete the circuit diagram for a relay that can latch a circuit.

Summary

A relay can be used to interface two circuits without an electrical connection or to create a latch.

Relays are NO before being switched.

Opto-isolators use an IR-emitting diode and a phototransistor to interface between electronic circuits.

kerboodle!

Examination-style questions

1. This question is about component identification and function.

(a) Provide the names and symbols of the components shown in Diagram **A**.

A

(10 × 1 mark)

AQA Examiner's tip Take care when drawing electronic circuits and symbols. If there is an error you will not be given the mark.

(b) Name an electronic component which best fits each of the descriptions given below:

(i) It allows current to flow in one direction only. *(1 mark)*

(ii) Its resistance decreases as the temperature increases. *(1 mark)*

(iii) It has two connections and stops a current flowing when it is pressed. *(1 mark)*

(iv) It has three connections called anode, cathode and gate. *(1 mark)*

(Total for Q1: 4 marks)

2. This question is about operational amplifiers and potential dividers.

A student wants to use a light sensor to act as an input trigger for a system that uses an LED.

(a) Diagram **B** shows the light sensor potential divider.

B

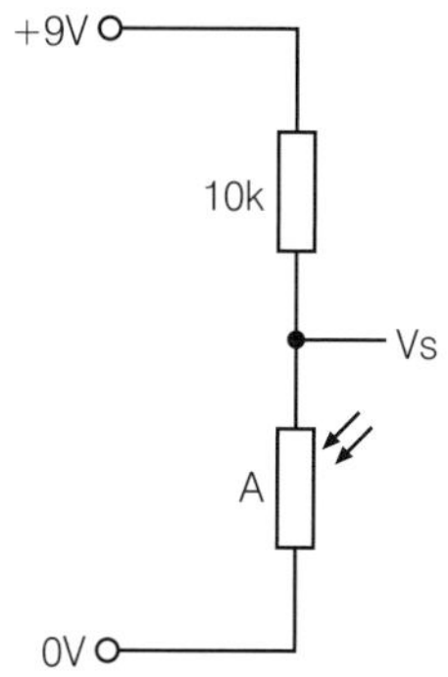

(i) Component A is a .... *(1 mark)*

(ii) In bright light conditions component A has a resistance of 5 k.
Calculate the value of the V_s output signal.

Formula: ....

Working: ....

Answer with units: *(4 marks)*

(iii) Explain what happens to the V_s output voltage when the light level falls. *(2 marks)*

(b) An operational amplifier can be connected as a comparator.

C

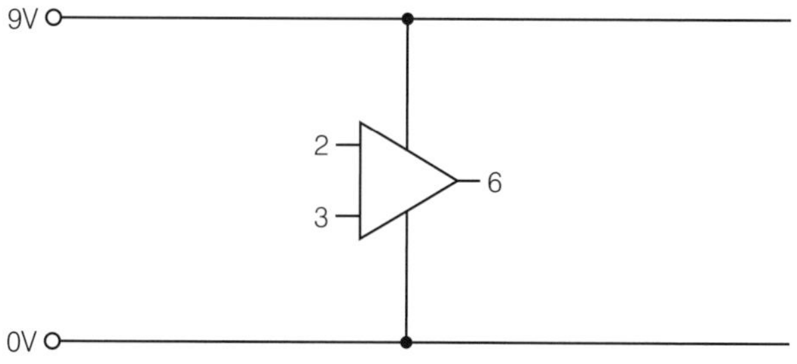

Complete Diagram **C** by adding:

(i) the light sensing potential divider from Figure 1.2 to the non-inverting input to give a positive output from the amplifier when the sensor is in the dark; *(2 marks)*

(ii) a potential divider connected to the inverting input to give an adjustable reference voltage; *(2 marks)*

(iii) an LED with a resistor to indicate that the output is high voltage. *(3 marks)*

(Total for Q2: 14 marks)

3. Diagram **D** shows a block diagram of a fire alarm.

D

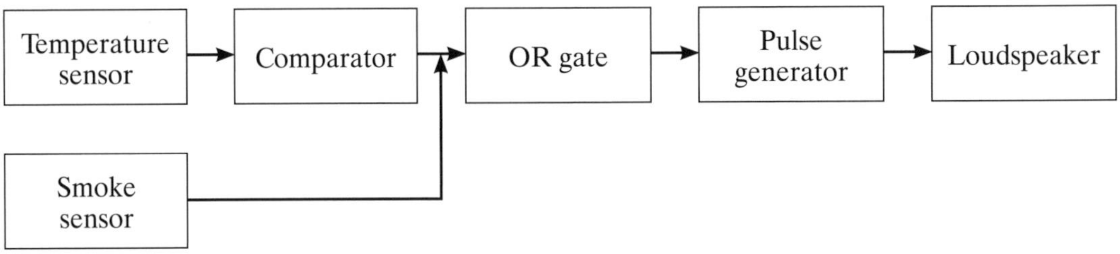

(a) State which block represents:

(i) the final output stage *(1 mark)*

(ii) an input stage *(1 mark)*

(iii) an astable. *(1 mark)*

(b) State the block in which you would find:

(i) an op-amp *(1 mark)*

(ii) a thermistor *(1 mark)*

(iii) the control of the frequency of the sound. *(1 mark)*

(c) Diagram **E** shows a pulse generator circuit used as part of the system.

E

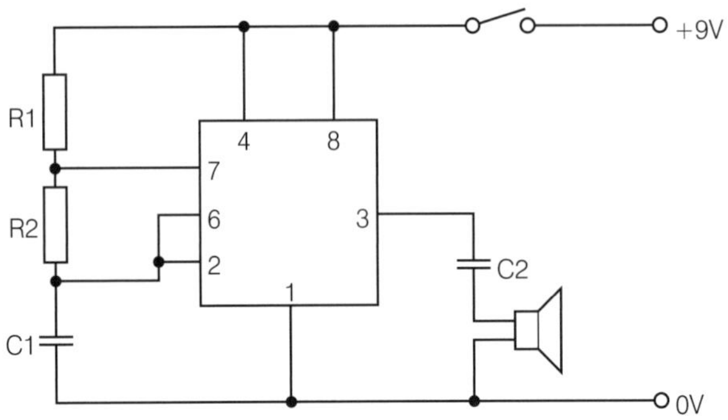

Component **C1** helps to control the frequency of the circuit.

(i) Which two components, other than **C1** in Diagram **E**, control the frequency of the circuit? *(2 marks)*

(ii) Explain the effect on the sound from the loudspeaker if the value of **C1** were increased. *(2 marks)*

(d) The final circuit could be constructed using either Veroboard (stripboard) or on a PCB.

Compare the use of Veroboard as opposed to a PCB for producing the circuit. *(2 marks)*

(Total for Q3: 12 marks)

4. This question is about PIC microcontrollers.

A student wants to make a rear lamp for a bicycle with four LEDs which flash in a pattern. This could be done using a PIC or by using a 555 timer circuit.

F LEDs

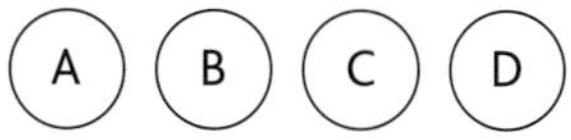

(a) Compare the use of a PIC with a 555 timer IC to control the LEDs. *(4 marks)*

(b) The student decides to use a PIC to make the LEDs flash on and off.

Using a programming system you are familiar with, produce a sequence of commands which would make a set of four LEDs, shown in Diagram **F**, switch on as shown below.

Next to each of the commands, explain its purpose.

(i) All four LEDs go high for 1 second.
(ii) A, B, C and D go high in sequence, each for 0.25 seconds.
(iii) All four LEDs go high for 1 second.
(iv) D, C, B and A go high in sequence, each for 0.25 seconds.
(v) This repeats continuously.
(vi) Between each statement, all LEDs go low for 0.25 seconds. *(8 marks)*

(Total for Q4: 12 marks)

UNIT ONE Written paper

What will you study in this section?

After completing Design, materials and manufacturing (chapters 5–7) you should have a good understanding of:

- design and market influences
- materials
- processes and manufacturing.

Design, materials and manufacturing

Designing and making complete products

Finished electronic products are normally made up of a circuit and an enclosure. The enclosure is the container or casing that is used to house the product. If a mobile phone were just a circuit board with the components attached, it would quickly get damaged and stop working. It might even present a safety risk to the user. The enclosure has important needs to fill for the finished product, including what it looks like.

In Systems and components we investigated the systems blocks used to design electronic circuits and the components they include. In this section we will investigate the design and manufacture of the complete product, including both the circuit and the enclosure. To put this in context, it is useful to have an overview of some of the things that need to be considered during the process of designing and making an electronic product.

Designing electronic products

It is a common mistake when designing an electronic product to think that the only important part is the circuit. There will be some things that the product is needed to do and it will be the job of the circuit to do these things. However, there will also be other needs that have to be met by the enclosure. For example, consider a personal MP3 player – it needs to be able to play music, but it also it needs to be small enough to carry, lightweight and, to sell well, it needs to look attractive to the user.

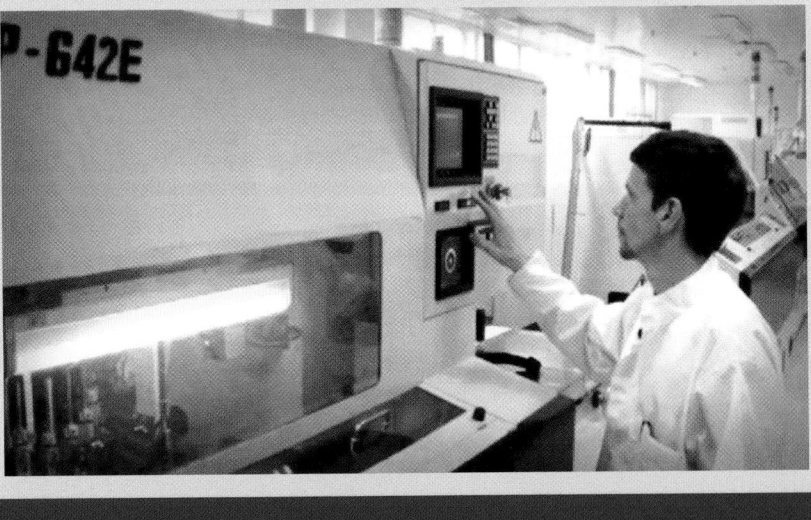

This means that the design of the circuit and the enclosure depend, in part, upon each other. The circuit may need to be designed to fit inside the enclosure. The enclosure may need to be tailored to meet the user's needs for the size of the finished product, whilst accommodating the size of the circuit.

When designing commercial products, there are even more design considerations. For example, what are the social, cultural and moral impacts of both the product and the manufacturing? What legal or standard requirements must the product meet? How will the different parts of the product affect the environment during its life and at the end of its life?

One essential task once the initial design has been developed is to select the materials that will be used to make the enclosure. There are many different materials, each with different qualities, strengths and weaknesses. Some have properties which can change in response to changing conditions in their surroundings. The properties of the materials have to be compared against the requirements of the design. In some cases, no material may be able to fill all the needs for a product to function as required, and it may be necessary to choose a material that is a 'best fit' for the need.

Manufacturing electronic products

When selecting materials, consideration has to be given to how the product can be made and what manufacturing processes are available. These will probably be affected by the quantity of products that are to be made. For example, for a one-off product, hand tools might be used to make the enclosure, but for a product being made in large quantities, the enclosure might be made using computer controlled machines. Similarly, for the circuit, there are several techniques to make one-off circuits by hand, but where large quantities of circuits are being made computer controlled equipment may be used to assemble and solder the circuits. Once the individual parts have been made, they then need to be put together as a finished product and tested to ensure that they meet the needs of the user.

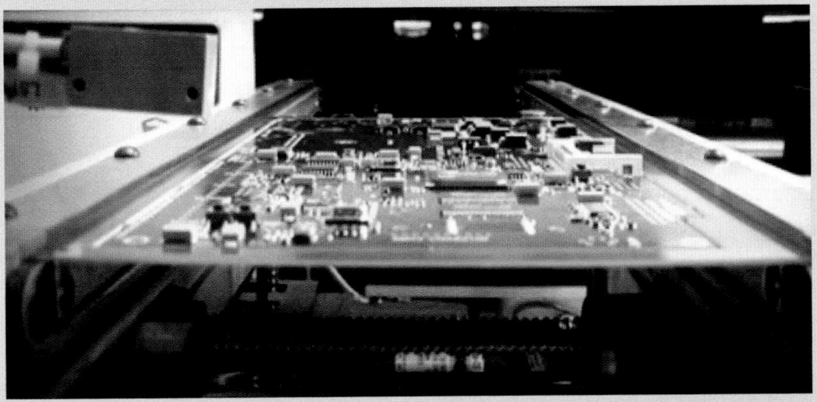

kerboodle!

5 Design and market influences

5.1 Design considerations

When designing a product, there are a wider range of factors to consider than just how well the product will work and what it will look like. Responsible design means having to make moral choices about a wide range of issues. These may be social, cultural, environmental or related to standards and regulations. Many of these choices will depend on market influences, such as the acceptable cost.

Objectives

- Explain that there are a broad range of social, cultural, environmental and cost issues that need to be considered during the design of a product.
- Describe how design for maintenance and recycling may impact product life.
- Give examples of how approaches to making products and using new products may affect society.
- Explain the role of standards when designing products.

Responsible design

Environmental and sustainability issues

Most products have a limited usable life. The length of this life, whether it can be extended and what happens to the product at the end of its life are important design considerations. This is sometimes called **planned obsolesence**.

Design for maintenance

Maintenance refers to any activity which extends the usable life of a product. This means that the product lasts longer, so that overall less material is needed to make replacements. Maintenance activities range from replacing batteries to repairing or replacing broken parts.

Consideration has to be given to whether it is more cost-effective or morally acceptable to repair an item or to replace it when it breaks down. A manufacturer may try to limit maintenance, with the intention of selling more replacement products. A more sustainable approach is to design products with access panels and standard screws, so that they can be repaired; or even to design products as a series of standard modules. This would mean that only the faulty module would need to be repaired or replaced.

Recycling

One design decision is whether a product will be capable of being **recycled**. For example, symbols are used on plastic products to show the type of plastic used, so that it can be sorted into different types and recycled, Figure A.

Electronic circuits are difficult to recycle, as the cost of removing the components currently exceeds the cost of replacing them. However, household appliances and computers can be taken to recycling centres. Care needs to be taken to dispose of hazardous chemicals in batteries and cells, and in other products such as refrigerators. If these were placed in landfill sites they would cause **pollution**.

Key terms

Planned obsolescence: designing a product for a limited life.

Maintenance: carrying out activities to extend the usable life of a product.

Recycled: made using materials that have been used before and reprocessed.

Pollution: contamination of the environment.

Automate: use computer-controlled machines instead of workers to perform tasks.

A *Example of a recycling mark on a plastic product*

Social and cultural issues

Making products

Everyone wants cheap, good-quality products. One decision that may reduce manufacturing costs is to **automate** manufacturing processes. However, this means that fewer people are employed.

An alternative way of reducing costs is to decide to make products in countries where labour costs are low. In these countries, sometimes the conditions for workers are far below the standards accepted in the UK. Pollution may also be higher. There is also an environmental cost in transporting goods all over the world.

The impact of electronic products on society

Electronic systems are part of our everyday lives. Products such as computers, televisions and mobile phones have a huge effect on the way that we live our lives and interact with each other.

For example, MP3 players are commonly used on public transport. They allow people personal choice of what they listen to. However, some people find the background noise from these devices irritating. They also argue that personal music systems may reduce the amount of communication between people. Others argue that they also reduce the possible markets for background music and advertising.

B *MP3 player*

C *British Standard Kitemark*

D *CE mark*

Standards and regulations

Standards exist to ensure that products meet suitable levels of safety and quality. In the UK, standards are regulated by the British Standards Institution (BSI). Products which meet these standards are awarded a British Standard and can be marked with the KiteMark, Figure **C**. Within the European Union products can also be awarded a CE mark, Figure **D**. This indicates to government officials that the product conforms to a standard which enables it to be legally placed on the market within their country.

Activity

Carry out a product analysis of an electronic product that is no longer in use (for example an old TV remote control). Investigate the materials used to make it. Are there any pollution or health issues with any of the materials used? Is it practical to separate the different materials? Is the material used to make the enclosure recyclable? Prepare a report on how it could be recycled.

Remember

Recycling and legislation

In January 2007, the Waste Electrical and Electronic Equipment Directive (WEEE Directive) came into force. This creates legal requirements for the disposal and recycling of electronic products.

Summary

Responsible design means having to make moral choices about social, cultural and environmental issues. Many of these choices will depend on market influences, such as the acceptable cost.

Design for maintenance and recycling can reduce environmental impact by reducing the need for new materials.

The choice of how and where a product is manufactured can affect society and culture.

Standards exist to ensure that products meet suitable levels of safety and quality.

AQA Examiner's tip

You should be able to identify a range of social, cultural and environmental issues that affect the design, manufacture and use of electronic products. Understanding one product in depth can help your understanding of other products.

kerboodle!

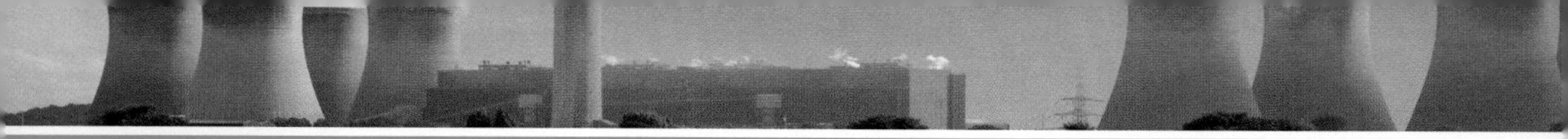

5.2 Computer-aided design

Designing a product can require great skill and a broad knowledge of materials, processes and electronics. Computer software is widely used to support these activities, speeding up the design process. Most people associate **computer-aided design (CAD)** with producing high quality working drawings. However, CAD software can be used to carry out many tasks during the development of electronic products, for both the circuit and the enclosure.

Using CAD when designing an electronic circuit

CAD software can be used throughout the process of circuit design. Its uses include:

- **modelling** the systems diagram for the circuit
- selecting the components to be used, creating the circuit diagram and modelling its operation. This is demonstrated in topic 8.6
- creating the 'real-world' circuit layout and the design of the **printed circuit board (PCB)**, Diagram **A**. This is explained in detail in topic 8.7
- for microcontroller circuits, creating the flowcharts or basic programs used to control the operation of the circuit, modelling how they work and downloading these into the microcontroller (see topic 3.14).

A major advantage of using CAD software is that it carries out any calculations required. This greatly reduces the amount of time needed to create a design and also gives confidence that the different components will work together. Many of the CAD software packages also allow the system or circuit to be modelled and tested virtually. This saves a lot of time and cost, by reducing the number of prototypes that need to be made.

Objectives

Describe how CAD software can be used to support the design of electronic products.

Key terms

Computer-aided design (CAD): the use of computer software to support the design of a product.

Modelling: simulating the use of a circuit or product.

Printed circuit board (PCB): the specially-designed base that an electronic circuit is assembled on using soldering.

Dimensions: sizes.

Features: details of the design.

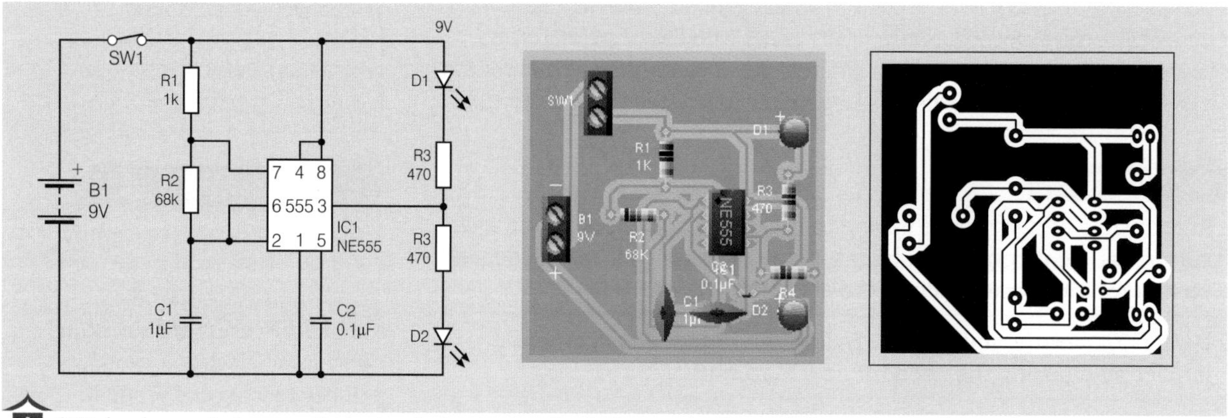

A CAD development of a monostable circuit, showing the circuit diagram, 'real-world' view and printed circuit board

Using CAD during the design of the enclosure

The main use of CAD software during the design of the enclosure is to produce drawings of the parts needed, showing the **dimensions**. CAD drawings have many advantages over drawing by hand:

- It is easier and quicker to make changes to a drawing. To make changes to a drawing made by hand, it is often necessary to restart the drawing from scratch. To make changes to a CAD drawing, you can open and edit the existing file.
- CAD drawings can be more accurate.
- CAD drawings can be saved electronically, saving space, and quickly circulated by email, saving paper.
- CAD drawings can be sent to computer controlled machines for the item to be made (see topic 7.4).

There are two main types of CAD drawing software: 2-dimensional (2D) and 3-dimensional (3D). Both types of software normally include a wide range of drawing tools. These are the basic commands used to create CAD drawings. The tools typically range from drawing simple lines and inserting shapes, through to duplicating and manipulating the drawn features and modifying and deleting features.

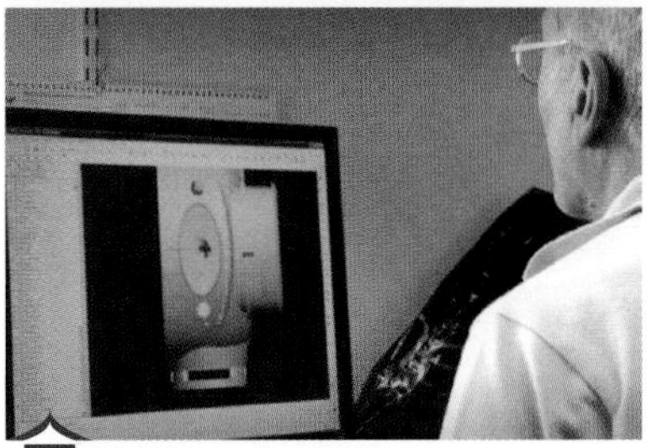

B *Creating a 3D CAD model*

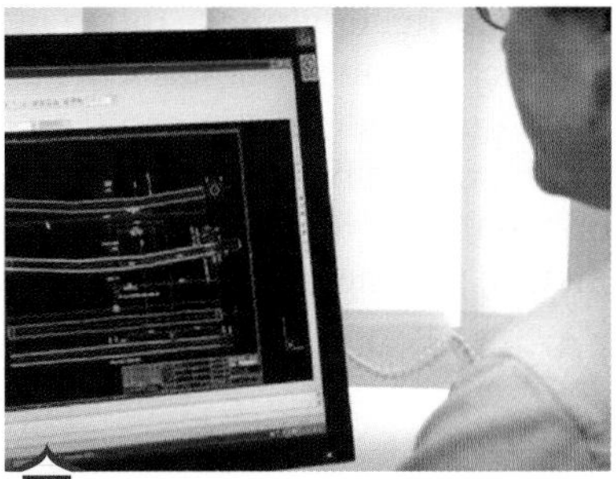

C *Creating a 2D CAD design*

2D CAD

Drawing using 2D CAD software is similar to drawing by hand. The screen has a working area, which is in effect the piece of paper that you draw on. This area can be changed to almost any size and you can zoom in to see features close up.

3D CAD

In 3D CAD, what is created is a three-dimensional model of the part being drawn, Diagram **B**. Starting to draw a 3D shape is more complicated than drawing a 2D shape. The design normally starts as a 2D drawing of the bottom of the shape. This is then extruded to give it thickness or depth, making it 3D. Other **features** can then be added to this design. The design can also be rotated and moved on the screen, so that it can be viewed and edited from any direction.

3D CAD drawings are especially useful for modelling the performance of a design. For enclosures, this can include checking that different parts fit together or carrying out stress analysis, which evaluates the strength of the design.

Summary

Computer-aided design (CAD) is the use of software to assist in the design of a product.

CAD software can be used to calculate which components to use in electronic products, and to create circuit diagrams and the real-world layouts of components.

CAD drawing software can be used to create 2D and 3D drawings of the enclosure.

CAD software can also be used to test virtual models of the circuit and enclosure, saving the time and cost of making a prototype.

Activity

Write a magazine article explaining how the end-user benefits from the use of CAD during the design of an electrical product of your choice. You should use a maximum of 200 words.

AQA Examiner's tip

You should be able to clearly explain at least two different examples of the use of CAD and explain the benefits of using CAD.

kerboodle!

6 Materials

6.1 Materials – wood and metal

Different materials have different properties. The material that a product is made from has a big effect on that product's ability to do what it is needed to do and the processes that are needed to make the product. Materials are normally grouped together into types based on what they were made from. This gives five main categories: wood, metals, polymers, ceramics and composites.

Recently, some materials have been developed with properties that can change in response to changes in their environment. Although each of these materials falls into one of the five types, they are sometimes referred to as a separate group called smart materials.

Objectives

List the main types of material.

Identify and describe the qualities of a range of wooden and metallic materials.

Key terms

Grain: the direction or pattern of fibres found in wood.

Softwood: wood from a coniferous tree.

Hardwood: wood from a deciduous tree.

Manufactured board: a wood product made by processing or pulping wood particles or sheets.

Alloy: a mixture of two or more metals.

Ferrous metal: a metal that contains iron.

Non-ferrous metal: a metal that does not contain iron.

Activity

Research as many different types of wood as you can. Make notes of their names, the regions where they are grown and typical uses.

Types of material

Wood and manufactured board

Solid timber

Solid timber is time-consuming to process and only comes in relatively narrow widths. Its properties may vary in different directions, depending upon its **grain**, Photo A. There are two types of solid timber: **softwood** and **hardwood**. These names do not refer to the properties of the wood: some softwoods, such as redwood, are hard and some hardwoods, such as balsa, can be soft.

Softwoods, such as pine, redwood and whitewood, come from coniferous trees that keep their leaves all year round. Softwoods can be grown in sustainable forests, meaning that more wood can be grown to replace that being used.

Hardwoods include oak, beech, elm, ash, mahogany, teak and balsa wood. These come from deciduous trees that lose their leaves each winter. They tend to grow more slowly than softwoods.

A *Photo of wooden chair that shows the grain in the wood*

Wood products

Manufactured boards are often made from the waste or off-cuts from cutting solid timber. They are available in large widths and their properties can be uniform in different directions. Some of the most common wood products are:

- Medium density fibreboard (MDF): this is made from wood pulp bonded with a polymer called urea formaldehyde.
- Plywood: this is made from sheets of veneer glued together at 90° to each other. As the grain is alternated, it is equally strong in both directions and will not split along the grain like solid timber.

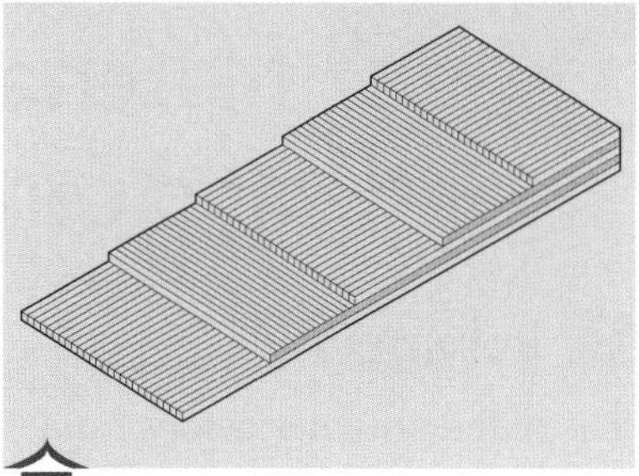

B *Structure of plywood*

Metals and their alloys

Metal is made from metal ore. This has to be mined and processed to turn it into a usable form. Metals are commonly available in a wide range of shapes and sizes. This reduces the amount of work that might be needed to change their shape. However, they are relatively difficult to work and join.

It is rare for a metal to be used in its pure form; normally they are mixed with other metals to improve their properties. A mixture of two or more metals is called an **alloy**. The proportion of the other metals added to form an alloy can typically range from 0.1 per cent up to 50 per cent.

There are two main types of metal alloy: **ferrous** and **non-ferrous**. Ferrous metals contain iron; non-ferrous metals do not contain iron. Both types of metal can be recycled.

The most common type of ferrous metal is low carbon steel, which contains up to 0.3 per cent carbon. Although it is the weakest of the steels, it is still stronger than most non-ferrous metals. Compared to other metals it is easy to machine, tough and costs less, but it can be prone to corrosion and rusting.

Most of the common non-ferrous metals have good corrosion resistance. These include:

- aluminium, which is lighter than steel, but more expensive and not as strong
- copper, which is a very good conductor of electricity. It is flexible, but lacks strength. It is commonly used to make electrical wires and pipework for domestic use, Photo C.
- zinc, which has a relatively low melting point, which makes it ideal for diecasting products such as camera bodies and handles for car doors.

C *Copper water pipes*

AQA Examiner's tip

When discussing the selection of materials, make sure that you describe the most important properties for the application, for example, weight and strength for making a portable enclosure.

Summary

Wood products are available as solid timber or manufactured board.

Softwoods include pine and redwood. Hardwoods include oak, beech, mahogany and balsa wood.

An alloy is a mixture of two or more metals. Ferrous metals are alloys containing iron. Non-ferrous metals do not contain iron.

Compared to other ferrous metals, low carbon steel is relatively cheap and has good strength. However, it is prone to corrosion.

Common non-ferrous metals include aluminium, copper and zinc.

kerboodle!

6.2 Materials – polymers, composites and ceramics

Polymers

The name **polymer** comes from the words 'poly', meaning many, and 'mer', meaning parts. They are manufactured by a process called polymerisation. This involves joining monomers together to form long chains of molecules. So, for example, polypropylene is made up of single monomers of propylene, joined together to form a long chain. A polymer product may contain a large number of the long chains.

There are two main types of polymer: thermosetting polymers, also called thermosets, and thermoplastic polymers, also called thermoplastics. Within these types, there are many different varieties of polymer with very different properties. Most types of polymer are good **insulators** against electricity and heat. Some of the stronger polymers compare favourably with the strength of metals. They are not normally painted, but their colour can be changed by adding pigments to them.

Thermosets

Thermosets are typically formed by a moulding process. During the process, they form many links across the different polymer chains, which stops the chains being able to move. This makes them harder and more rigid than thermoplastics, with good resistance to electricity and heat, Photo **A**. Once moulded, they cannot normally be reshaped and they cannot be recycled. Common thermosets include: phenol formaldehyde, used to make light sockets, urea formaldehyde, used to make light switches and epoxy resin, used to make printed circuit boards.

Thermoplastics

Thermoplastics do not have links between the different polymer chains. This means that they are softer and more flexible than thermosets. Thermoplastics soften when heated and can be shaped when hot. The shape will harden when it is cooled, but can be reshaped when heated up again. Thermoplastics can normally be recycled. They are readily available in sheets of standard thicknesses or in granular forms for use with moulding processes. There is a wide range of thermoplastics in common use.

Ceramics

Ceramic materials are often oxides, nitrides or carbides of metals. They have excellent corrosion resistance, generally have good strength in compression and are harder than most other engineering materials. This means that they are very resistant to scratches and wear. However, they tend to be weak in tension and brittle – this means that if they are pulled they don't stretch, but tend to crack and fall apart.

Ceramics are excellent insulators against both heat and electricity. They can normally withstand high temperatures without softening. They are used rather than polymers in applications where electrical insulators are needed and where flexibility is not required.

Objectives

Identify and describe the qualities of a range of polymer, composite and ceramic materials.

Key terms

Polymer: an organic material made up of a chain of single units called monomers.

Insulator: a material that does not allow heat or electricity to pass freely through it.

Ceramic: an inorganic material, normally an oxide, nitride or carbide of a metal.

Composite: a material that is made from two or more material types that are not chemically joined.

AQA Examiner's tip

You should not use generic names such as plastic or metal when selecting materials for your project. You should use the names of individual materials, for example HIPS, acrylic or low carbon steel.

A *Example of a product made from thermosets: plug and socket*

B *Typical uses of some common thermoplastics*

Thermoplastic	Properties	Examples of what it is used for
Polypropylene	Softens at 150°C. High impact strength for a polymer. Can be flexed many times without breaking.	Food containers, plastic chairs, children's toys
High impact polystyrene (HIPS)	Softens at 95°C. Easy to mould. Light but strong.	Vacuum-formed packaging and casings
Acrylic (Polymethyl methacrylate – PMMA)	Good optical properties. It can be transparent, like glass, or coloured with pigments. Hard-wearing and will not shatter on impact.	Machine guards, plastic windows, bath tubs, display signs
Nylon	Low friction qualities. Good resistance to wear.	Bearings, gear wheels, curtain rail fittings
High density polyethylene (HDPE)	Softens at 120°C. Strong.	Bowls, buckets, milk crates
Low density polyethylene (LDPE)	Softens at 85°C. Softer, more flexible and less strong than HDPE.	Detergent bottles, carrier bags, packaging, film
Polyvinyl chloride	Stiff and hard-wearing. Can be made softer and rubbery by adding a plasticiser.	Chemical tanks, pipework, coverings for electric cables, floor and wall coverings, packaging

Ceramic components are difficult to machine due to their hardness and the risk of breaking. They are frequently made by moulding processes, often using high temperatures, so that machining is not required. Whilst many ceramic materials could be recycled in principle, with the exception of glass it is not normally cost-effective to recycle ceramic materials.

Composites

A **composite** is a material that is made by combining two or more different types of material. The materials are not joined chemically – they still remain physically distinct. Composites combine the properties of the materials that they are made from.

Activities

1. Some plastics are marked with a symbol containing a number, to show that they can be recycled. The number identifies the type of polymer. Find out what the different numbers mean and identify an example of a product made with each of the different types of polymer that can be recycled. An image as an example may aid understanding.
2. Recommend a suitable polymer to make the cover for an MP3 player. Provide an explanation for your choice, by comparing it to other polymers.

Summary

Thermosetting polymers cannot be reshaped or recycled. They are strong and good insulators.

Common thermoplastics include HIPS, polypropylene and acrylic. These can be softened and reshaped when heated.

Ceramic materials have excellent corrosion resistance, and are good insulators. Although they are amongst the hardest engineering materials, they tend to be weak in tension and brittle.

Composites are a combination of two or more different types of material. They combine the properties of the materials that they are made from.

Remember

Uses of ceramics

Electrical insulators, such as the alumina casings on spark plugs

Tools for grinding and cutting

Tiles to insulate furnaces

Glass lenses

Building materials, such as plaster, cement and bricks.

Remember

Examples of composites

Material:	Example of its use:
Reinforced concrete	Buildings and construction
Fibreglass	PCBs, canoes, vehicle body panels
Carbon Reinforced Polymer (CRP)	Racing car bodies, helmets, armour
Metal matrix composites	Drive shafts and cylinders in high-performance cars

kerboodle!

6.3 Smart materials

Smart materials have properties that can change in response to changes in their environment. This means that they have one or more properties that can be changed by an external condition, such as temperature, light, stress or electricity. This presents the designer with some exciting options to consider for new products in the future. Some of the smart materials that are already finding uses include:

- **shape memory alloys**
- **piezoelectric materials**
- quantum tunnelling composite
- colour change materials.

Shape memory alloys

Most materials show some, but limited, memory due to elasticity – when stretched a little, they can spring back to their old shape. However, when stretched further or bent, they stay that way. If a part made from a shape memory alloy (SMA) is bent out of shape, when it is heated above what is known as its transition temperature, it will return to its original shape. This cycle of bending and being straightened can be repeated many times. SMAs can be formed into almost any shape, from springs to flat plates, and be conditioned to return to this shape by heating them above their transition temperature.

The most common SMA is an alloy of the metals nickel and titanium, which has a transition temperature of 70°C. The heating can be achieved through direct means or by passing an electric current through it.

Piezoelectric materials

Piezoelectric materials do not conduct electric current. However, when squeezed rapidly they produce an electrical voltage for a moment. Alternatively, if a voltage is put across the material it creates a tiny change in shape.

An example of a natural piezoelectric material is quartz. A wide range of materials with piezoelectric behaviour have been developed, including polymers and thin film ceramics. These have found many uses, including:

- contact sensors for alarm systems
- microphones and loudspeakers
- electrical generators
- motors to move the lenses on cameras
- regulators for electric circuits, such as quartz clocks.

Objectives

Explain what is meant by smart material.

Describe the properties and typical uses of a range of smart materials.

Remember

Applications of shape memory alloys

Triggers to start the sprinklers in fire alarm systems

Controllers for hot water valves in showers or coffee machines

Shrink-fit seals for hydraulic tubing

Artificial muscles in robot hands

Spectacle frames.

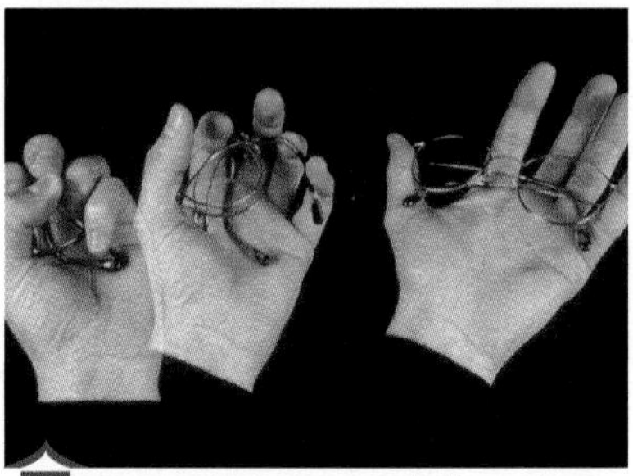

A *Spectacle frames made from shape memory alloy*

Key terms

Shape memory alloy: a metal that, once deformed, will return to its original shape when heated above its transition temperature.

Piezoelectric material: a material which changes shape fractionally when a voltage is applied to it.

Thermochromic: changes colour with temperature.

Photochromic: changes colour according to different lighting conditions.

Quantum tunnelling composite

Quantum tunnelling composite (QTC) is a flexible polymer which contains tiny metal particles. In its normal state it is an insulator, but when squeezed it becomes a conductor able to pass high currents. QTC can be used to make membrane switches like those used on mobile phones, pressure sensors and speed controllers.

Colour change materials

Thermochromic materials change colour as the temperature changes. The colour changes are based on liquid crystal technology. At specific temperatures the liquid crystals, which may be as small as 10 microns in diameter, re-orientate their structure to produce an apparent change of colour.

Examples of the use of these materials include:

- plastic strips that use colour changes to indicate temperature or act as thermometers
- test strips on the side of batteries. These heat a resistor printed under the thermochromic film. The heat from the resistor causes the film to change colour
- packaging materials that show you when the product they contain is cooked to the right temperature
- colour indicators on cups, to show whether the contents are hot.

Photochromic materials change colour according to different lighting conditions. They are particularly reactive to ultraviolet light. They are used for products ranging from nail varnish, to security markers that can only be seen in ultraviolet light, jewellery and mobile phone cases.

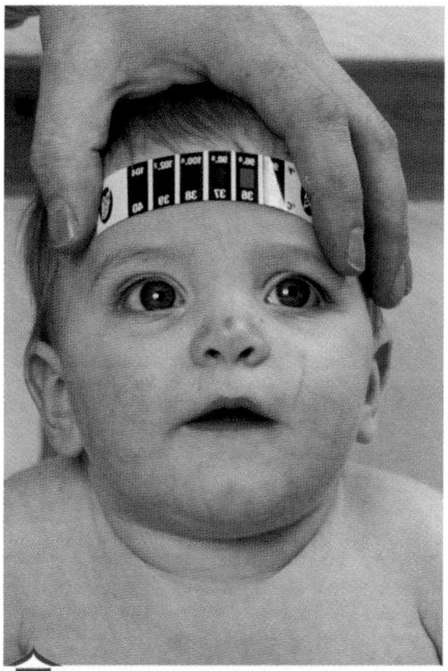

B *Contact thermometer made from thermochromic sheet*

Activities

1. Smart materials are being developed which can change properties when subjected to a magnetic field. Identify what these materials are made from and their potential applications.
2. Choose one type of smart material. Write a short report explaining which other materials it could be used to replace, explaining its potential advantages and drawbacks.

AQA Examiner's tip

Use real-life examples when describing the use of new and smart materials. Say why they have an advantage over traditional materials.

Summary

Smart materials can change one or more of their properties in response to changes in their environment.

There are smart materials which can change shape with temperature, change electrical resistance with stress or change colour with temperature or light levels.

6.4 Selecting the materials to make an enclosure

In topics 6.1, 6.2 and 6.3 we investigated the characteristics of different types of material. The reason for doing this was to allow informed choices to be made when selecting the materials that can be used to make the enclosure for an electronic product.

Material selection

The properties of the material used to make an enclosure must meet the needs of the product. For example, the enclosure for an alarm that will be used outside might need to be waterproof, but an alarm used indoors might not need to be. Based upon these needs, the important properties needed from the material can be identified. Examples of some design needs and the relevant important material properties are shown in Table **A**.

A *Considering what material properties the design needs*

Design question	Material property
Does the enclosure have to prevent electricity passing through it?	Electrical conductivity
Does the enclosure have to withstand forces, such as loads being put on it?	Strength
Does the enclosure need to be resistant to scratches and wear?	Hardness
Does the material need to be resistant to knocks and bumps?	Toughness
Does the enclosure have to be light, so that it can be moved easily or carried around?	Weight
Does the enclosure need to be in a certain price range?	Cost
Does the enclosure have to work in an environment that could damage it?	Corrosion resistance
Should the enclosure be a particular colour, texture or style?	Aesthetics
Does the enclosure need to be recycled?	Sustainability
Does the material need to stop heat from passing through it?	Thermal conductivity

Chooser charts

Table **B** compares some of the properties of a range of different materials. This is a subjective comparison and the ratings will vary for different types of application. A chooser chart like this can help to identify the materials with the combination of properties that may satisfy the needs of the application. However, in drawing up a shortlist, it will also be necessary to consider aesthetic issues, such as what the material looks like, and manufacturing issues such as materials availability and the ability of the available equipment to process each material. Once the possible materials have been identified, the list of those that are suitable should be tested to see how well they meet the needs of the product.

Objectives

List properties that might be considered when selecting a material to be used for an enclosure.

Explain how a chooser chart can be used to assist the selection of materials for an application.

AQA Examiner's tip

You should be able to compare the properties of different materials using several parameters, for example whether they are insulators or conductors, their strength, cost etc.

Activity

For each of the following products, identify the material properties that are needed and recommend a suitable material. You will need to provide an explanation for your choices, by comparing your recommendations to other materials:

- a one-off casing for a robot buggy
- the 'tamper-proof' enclosure for an alarm used on mountain bikes, made in batches of 10
- the case for the remote control of a new range of televisions, to be made in large quantities.

B *Subjective comparison of material properties*

Material	Strength	Hardness	Toughness	Weight	Corrosion resistance	Ability to conduct electricity	Ability to conduct heat	Ability to be recycled	Cost
	Comparative properties								
Pine (softwood)	Medium/ low	Low	Low	Low/ medium	Poor	Very poor	Very poor	Poor	Low
Beech (hardwood)	Medium	Low/ medium	Low/ medium	Low/ medium	Poor	Very poor	Very poor	Poor	Medium
Plywood	Low	Low	Low	Low	Poor	Very poor	Very poor	Poor	Low
Low carbon steel	Very good	Good	Very good	High	Poor	Very good	Very good	Excellent	Low
Stainless steel	Excellent	Very good	Very good	High	Good	Very good	Very good	Excellent	High
Aluminium alloys	Very good	Good	Very good	Low	Very good	Very good	Very good	Excellent	Medium
Zinc	Good	Good	Very good	Medium/ high	Good	Very good	Very good	Excellent	Medium
Copper	Medium	Medium	Good	High	OK	Excellent	Excellent	Excellent	High
Brass	Good	Good	Very good	High	Good	Very good	Very good	Excellent	High
Melamine formaldehyde	Medium	Very good	Low	Medium	Very good	Very poor	Very poor	Very poor	Medium
Phenol formaldehyde	Medium	Good	Low	Medium	Very good	Very poor	Very poor	Very poor	Medium
Polypropylene	Low/ medium	Low	Good	Low	Very good	Very poor	Very poor	Excellent	Low
High impact polystyrene (HIPS)	Medium	Low	Good	Low	Very good	Very poor	Very poor	Very good	Low
Acrylic (PMMA)	Medium	Low/ medium	Good	Low	Good	Very poor	Very poor	Very good	Medium
Nylon	Good	Good	Good	Low	Very good	Very poor	Very poor	Very good	Low
High density polyethylene (HDPE)	Low/ medium	Low	Good	Low	Very good	Very poor	Very poor	Excellent	Low
Low density polyethylene (LDPE)	Low	Low	Good	Low	Very good	Very poor	Very poor	Excellent	Low
Polyvinyl chloride	Low	Good	Medium	Low	Very good	Poor	Very poor	Good	Low
Alumina	Very good	Excellent	Poor	Medium	Excellent	Very poor	Very poor	Low	High
Alumina silicates	Good	Excellent	Poor	Medium	Excellent	Very poor	Very poor	Low	High
Fibreglass	Very good	Good	Very good	Low	Good	Very poor	Poor	Very poor	High
Carbon reinforced plastic (CRP)	Very good	Medium	Excellent	Low	Very good	Poor	Poor	Very poor	Very high
Shape memory alloy	Very good	Good	Very good	Medium	Good	Very good	Very good	Very good	High

Summary

Different materials have a wide range of properties and different combinations of properties.

A chooser chart is a useful tool to help select materials that could be used to make a product. However, the designer will still have to evaluate the properties of these materials against the needs of the product using quantitative data and will need to consider manufacturing issues.

kerboodle!

7 Processes and manufacturing

7.1 Types of production

The quantity of a product that needs to be made has a significant effect on how the products are manufactured, Table **C**. There are three categories used in GCSE Electronic Products for the types of production.

Objectives

- List three categories or types of production.
- Explain how the approaches used to manufacture electronic products vary with the type of production.
- Explain the difference between quality assurance and quality control.
- Explain how the type of production affects the approach used for managing quality.

One-off production

One-off production is making a single item, usually to a customer's individual specification. One-off electronic products include a sound system for a theatre and a control panel used on a spacecraft.

One-off products are normally made by skilled workers, either by hand or using standard workshop equipment. The product costs are very high because of the amount of labour time needed.

A *A satellite is an example of a one-off product*

Batch production

If a company uses **batch production** to make 300 alarms, it doesn't make them one after the other, like one-off production. It makes 300 circuits, 300 enclosures, and then puts the parts together. Batch production is used to make quantities of products ranging from a few to a few thousand. Often these products have similar designs, but are customised in some way. Examples include electronic speedometers used in Formula One cars or amplifiers for electric guitars.

The equipment used for batch production must be flexible, so that it can quickly change to making a different product. Workers must be skilled to make these changes. The labour cost is lower than one-off production, as making similar parts together needs fewer changes to machines.

B *Electric guitars are an example of batch production*

High-volume production

Products manufactured in **high-volume production** include cars, mobile phones and televisions. Most things that you experience every day will probably be mass-produced.

The cost of creating a production line for high-volume production is normally very high. This means that large numbers of products have to be made so that the cost of the equipment can be divided between them. Most of the machines used will be designed to just do one job, again and again. Often the human workforce is relatively unskilled because a lot of the manufacturing processes used are computer-controlled.

AQA Examiner's tip

You should be able to give examples of products made using the different types of production.

How the type of production affects quality assurance and quality control

Quality assurance (QA) means taking steps **before making** a product, to make sure that the manufacturing process makes the product correctly. **Quality control (QC)** means checking a product **after it is made** to see if it was made correctly.

In one-off production, the emphasis is normally on QC. A worker will make a part and check after each process to see whether it is correct. If it is wrong, he will modify it to make it right.

In high-volume production this method would cost too much. The emphasis is normally on QA. The manufacturing processes are tightly controlled, so that the products are always made to the requirements. This may involve putting sensors on machines or using work-holding devices to make sure that the products are always in the same place. Normally, only a few parts will be checked after each process, just to make sure that the process is running correctly.

Batch production will combine aspects of both these approaches, depending upon the number of parts to be made.

C *Characteristics of different scales of production*

	One-off production	Batch production	High volume production
Number of products to make	lowest ←	←	highest
Cost of equipment	lowest ←	←	highest
Cost of labour per product	highest →	→	lowest
Product cost	highest →	→	lowest
Typical type of equipment used to make the circuit	Manual soldering	Manual soldering, wave soldering	Pick and place machines, reflow soldering
Typical type of equipment used to make the enclosure	Hand tools and simple machine tools	Flexible machines	Dedicated computer-controlled machines

Note: the processes and equipment used will be explained later in this chapter

Summary

The type of production – one-off, batch or high-volume – affects the method used to manufacture a product. As the number of parts to be made increases, there is normally increased use of computer-controlled machines.

Quality assurance means taking steps before making a product to ensure that it is made correctly. Quality control means checking a product after it has been made to see whether it is correct.

As the number of parts to be made increases, the emphasis for managing product quality normally moves from QC to increasing QA.

Key terms

One-off production: making a single product or prototype.

Batch production: making a quantity of parts before switching to making another product.

High-volume production: making large numbers of parts using dedicated machines.

Quality assurance (QA): taking steps before making a product to ensure sure that it is made correctly.

Quality control (QC): checking that a part or product is correct after it has been made.

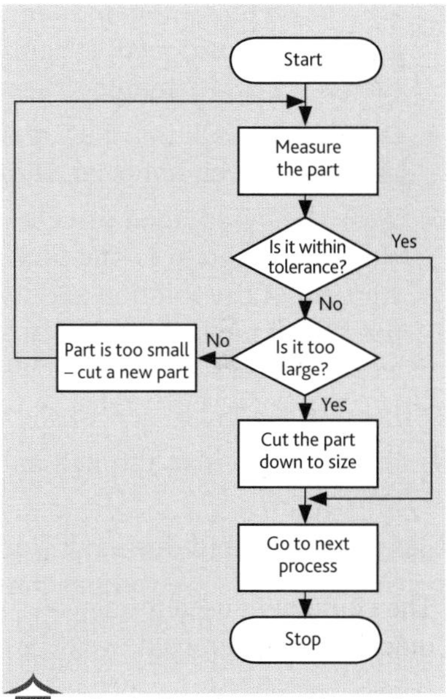

D *A simple flowchart used for quality control during the cutting of a steel sheet*

Activity

Find five examples of engineered products made using each of the three types of production.

7.2 Making PCBs

Circuits are normally assembled on a printed circuit board (PCB), for example Diagram **A**. The material used to make a PCB consists of a layer of thin copper bonded to a sheet of **fibreglass**. The finished PCB will have the **tracks** that link the components on it, with **pads** where the components are attached. The copper around this layout will be removed; most often, this is carried out by a process called photo-etching. However, circuit boards can also be produced using a computer-controlled router.

Objectives

Describe how to produce a PCB using the photo-etching process.

Describe how to produce a PCB using a computer-controlled router.

Key terms

Fibreglass: a non-conductive composite material made from glass fibres and plastic resin.

Track: a 'path' of copper, joining components.

Pad: a contact point for a component.

Mask: a pattern used to shield areas of the photosensitive PCB from light.

AQA Examiner's tip

You should be able to explain all the safety precautions needed during the photo-etching process.

Making PCBs using the photo-etching process

This process uses PCB materials coated with a photosensitive film, Diagram **B**. The steps involved are:

1. Design the PCB layout using CAD software, Diagram **C**. This is demonstrated in topic 8.7.
2. Print the layout onto an acetate sheet using a laser printer. This is called the PCB **mask**.
3. Remove the protective plastic cover on the photosensitive PCB. Place it on top of the mask in a lightbox, ensuring that the mask is correctly positioned, Diagram **D**. Expose to ultraviolet light for between 3 and 7 minutes, depending upon the equipment used.
4. Develop the PCB in a bath of developer solution until the exposed areas have been removed, Diagram **E**.
5. Wash the board, then place it into an etch tank containing ferric chloride, Diagram **F**. The time that this takes will depend upon the strength of the solution and its temperature. Too little time will not fully remove the exposed layer. Too much time, and the non-exposed tracks may also be washed off.
6. Remove the board and wash thoroughly.
7. Drill holes so that 'through hole' components can be positioned, Diagram **G**.

The chemicals used during this process are highly corrosive and harmful. Suitable precautions must be used, such as handling the PCB with tongs, and wearing gloves, goggles and aprons. This process produces a considerable amount of corrosive waste material which requires careful disposal.

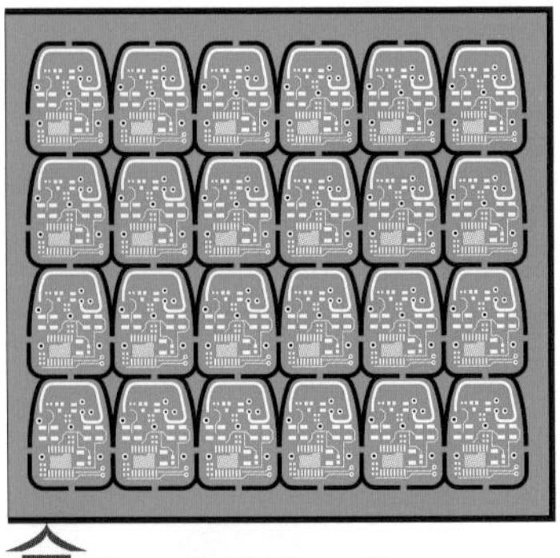

A *Shaped PCBs for car keys*

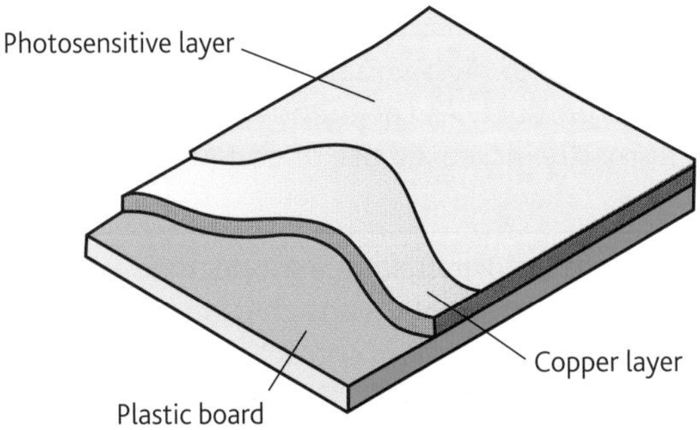

B *PCB board for photo-etch process*

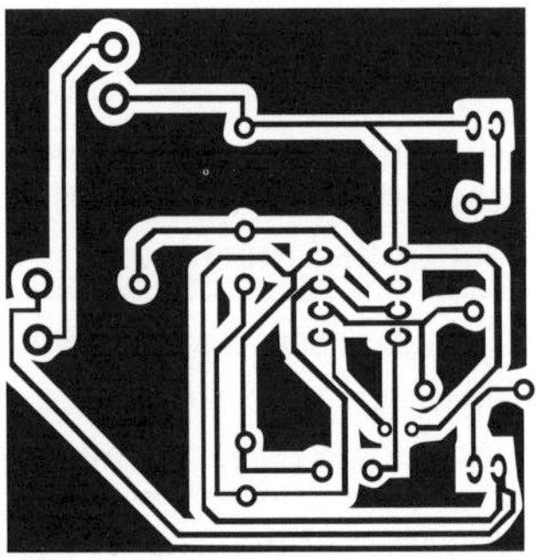

C *PCB design*

D *Lightbox*

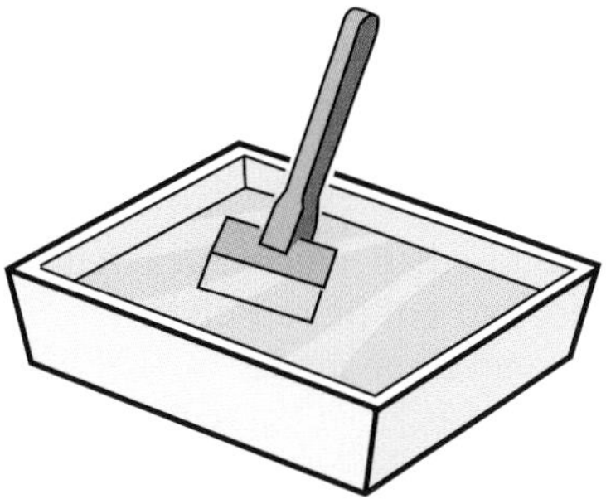

E *Developing the exposed board*

F *Etching tank*

Making PCBs using a computer-controlled router

One-off PCBs are often made using computer-controlled routers and milling machines. The circuit board used in this process does not need a photosensitive coating. The machine removes the copper around the tracks and pads.

The instructions for operating the router are fairly straightforward, but step 2 requires many calculations:

1 Set the tool movement speed.
2 Enter the sequence of movements needed. These will start from a start point called the datum and follow each of the areas where the copper is to be removed.
3 Turn the chuck (tool) on.
4 Carry out the tool or job movement.
5 Return the tool to the start point.
6 Turn chuck off.

The machine is normally controlled by computer software. It takes the PCB layout directly from the CAD software and turns this into a string of numbers which tell the router where to move. For one-off PCBs, this approach is more energy-efficient than photo-etching and produces fewer harmful waste products.

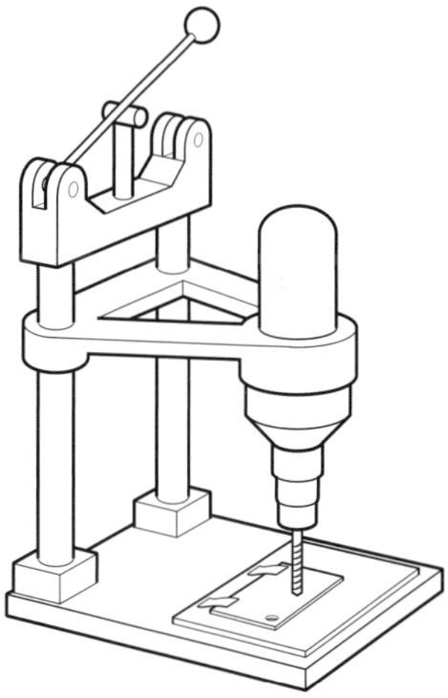

G *PCB drill*

Activity

Produce a flowchart with quality control checks and safety notes which illustrates the stages involved in the production of a printed circuit board.

Summary

Photo-etching is used to produce PCBs using a mask of the PCB design, a lightbox and a series of different chemicals. Safety precautions must be taken when using the chemicals.

Computer-controlled routers can produce circuit boards by removing the material around the tracks and pads.

kerboodle!

7.3 Assembling circuits

Making circuits

There are several methods that can be used to create electronic circuits. Selecting the most suitable method will depend on the reason for making the circuit and how many circuits need to be made.

Prototyping kits

Prototyping kits consist of sets of systems blocks. The user selects suitable input, process and output blocks needed and plugs them together, Photo **A**. These are used to develop or test circuit ideas. They are quick to use and to modify, but limit the flexibility of design and can need a lot of space.

Prototype boards

A prototype board, also known as a **breadboard**, is a matrix board consisting of holes into which individual components can be inserted. The holes are joined together in rows in order to achieve electrical continuity, Photo **B**. These are used to make one-off circuits, to develop or evaluate circuit ideas. Components can be easily changed or removed without damage and the circuit can be reliably tested without soldering.

Stripboard

Stripboard is made from a polymer or composite base coated with strips of copper. Each strip of copper has holes predrilled at regular intervals. Components can be inserted through the board and soldered to the copper strips. The circuits are much more robust than breadboards. Stripboard is particularly useful for making small numbers of circuits when the facilities to manufacture PCBs are not available.

PCBs

Methods of PCB manufacture were discussed in topic 7.2. PCBs are normally the most reliable and space-efficient method of creating a circuit. For low-volume production, PCBs normally include holes that the components are located through – this is called the **through hole** technique. When manufacturing in large quantities, it can be time-consuming to drill holes and difficult to locate components in them. An alternative is to position and fix the components to the surface of the board, called the **surface mount** technique. Surface mount also allows both sides of a board to be used, saving space.

Populating PCBs

Populating means putting the components on the PCB. It is important that components are inserted with the correct orientation. For example, diodes and ICs need to be the correct way round. When manufacturing in small quantities, components are normally placed on the PCB by hand. When large quantities are being made, the components are normally positioned using computer-controlled machines (see topic 7.4).

Objectives

- Describe how circuits can be created using a range of different approaches.
- Explain how the volume of production affects the technologies used to make circuits using PCBs.

Key terms

Breadboard: a commonly used name for a prototype board.

Stripboard: a polymer or composite board coated with strips of copper, with holes at predrilled intervals.

Through hole: an approach where components need to be put through holes in the PCB.

Surface mount: an approach where components are attached to the PCB surface.

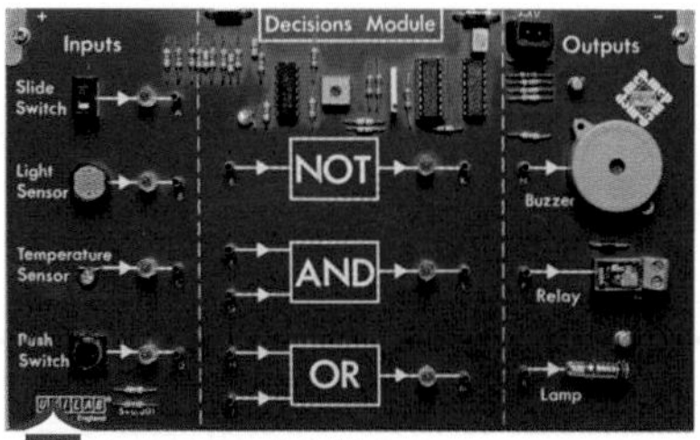

A *A prototyping kit*

Attaching components to the PCB

Manual soldering is used to make circuits in small quantities, Photo **C**. The safety issues to be considered will be covered in topic 8.11. It is important to ensure that a good joint is formed, Diagram **D**. A dry joint may result in intermittent operation or failure of the circuit to work.

Wave soldering is used to solder components onto large circuit boards during batch or large-scale production. It is suitable for use with either through-hole or surface mount components. The assembled PCB is clamped to a conveyor which passes over a wave of molten solder. The solder just touches each component and securely attaches it to the board. A complete PCB can be soldered in seconds.

Reflow soldering is the most widely used method of PCB production. It is particularly useful when working with small components. Solder paste is applied to the board and the components are placed in position. The board is then heated, either by an infra-red lamp or in a conventional oven. The solder melts and the components are joined to the board. Surface tension within the solder aligns any components that have not been accurately placed. This has helped to reduce the size of many electronic products, with considerable cost reduction.

(a) (b)

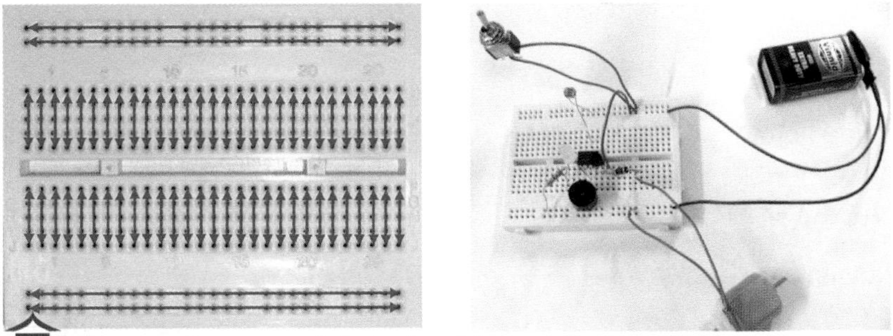

B ***a*** *the connections on a prototyping board;* ***b*** *a prototyping board in use*

Remember

Safety rules for soldering

Always treat a soldering iron as if it is hot.

When not being used, always put the soldering iron in its stand.

Only one person can use a soldering iron at a time.

Always use a baseboard.

Never use a piece of solder shorter than your little finger.

Never breathe in the smoke from the solder.

Never touch anything with the hot end except for the purpose of soldering.

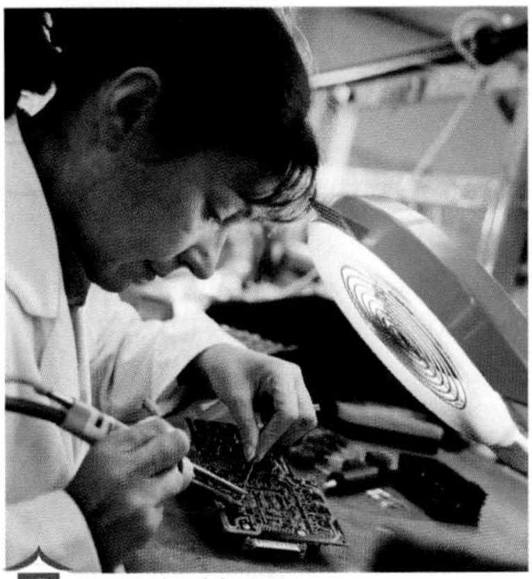

C *Manual soldering*

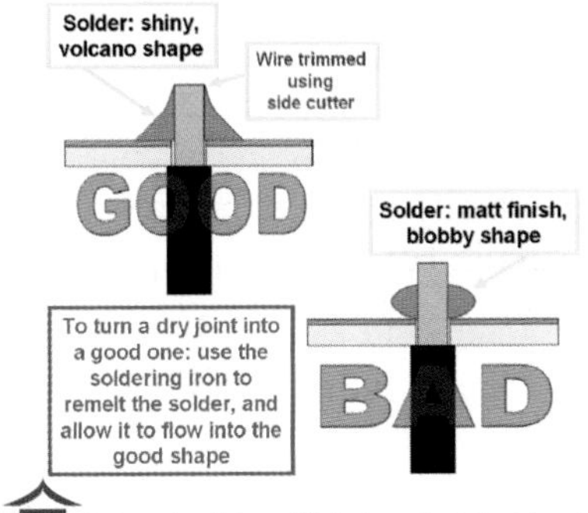
Solder: shiny, volcano shape
Wire trimmed using side cutter
GOOD
Solder: matt finish, blobby shape
To turn a dry joint into a good one: use the soldering iron to remelt the solder, and allow it to flow into the good shape
BAD

D *A good soldered joint and a 'dry' joint*

Activity

Carry out a product analysis of an electronic product, such as a computer board. Identify whether it uses through-hole or surface mount technology. Measure the sizes of some of the components and examine how accurate the soldering is.

Summary

Prototyping kits, prototype boards (breadboards) and stripboard can be used to produce prototype circuits.

PCBs using through hole components are typically used to make small quantities or batches of circuits. These may be soldered by hand or by wave soldering, depending upon the number of circuits to be made.

When making very large quantities of circuits, surface mount technology is commonly used, with reflow soldering.

AQA Examiner's tip

Make sure that you can outline the advantages of surface mount technology when producing printed circuit boards.

kerboodle!

7.4 Computer-aided manufacture

What is CAM?

Computer-aided manufacture (CAM) means using computers to operate machines to produce a product. CAM is widely used across all sectors of industry. All machines used for CAM use **computer numerical control** (CNC).

What is CNC?

CNC means using numerical data to control a machine. All machines that are controlled by a computer are CNC because all machine code is numerical.

The instructions to operate most machines are fairly straightforward. For example, they will turn the machine on, tell it what speed to move at, give it a series of movements to carry out and then turn the machine off. Similar instructions are enough to operate a lathe, milling machine, router, drill or other devices. These instructions can be turned into a program that will allow a computer to control a machine, rather than a skilled worker. The computer will read this program as strings of numbers. An example of an actual program is shown in Photo **A**.

A *Program on a CNC centre lathe*

What are the advantages of CAM?

CAM using CNC machines has several advantages over manual machines and human labour:

- The machining speeds are typically higher, so the parts are machined in less time.
- CNC machines do not need to take breaks – they can work continually, 24 hours a day, seven days a week, if needed.
- They are normally more accurate.
- They offer greater consistency, which can otherwise be a challenge when manufacturing products in large numbers.
- CNC mills can machine shapes that are difficult to manufacture using manual machines, such as freeform curves or the tracks on a circuit board.

There are two main disadvantages to using CNC machines:

- CNC machines are much more expensive than manual machines.
- It can take a long time to write the program to operate the CNC machine.

Objectives

Explain what is meant by CAM, CNC and CIM.

Describe the advantages and disadvantages of using CNC machines for high-volume production.

Key terms

Computer-aided manufacture (CAM): using computers to operate machines to produce a product.

Computer numerical control (CNC): using numerical data to control a machine.

Computer integrated manufacture (CIM): computer integrated manufacture, the use of CAD to design a product which is then transferred directly to be made on a CAM machine.

Types of CNC machine

The use of CNC machines is very widespread. Even within a school workshop, common CNC machines include sticker cutters, routers and mills, lathes and laser cutters.

Making circuit boards and circuits

CNC routers can be used to make printed circuit boards for one-off products (see topic 7.2).

An important use of CNC machines in high-volume production is to put the components in the correct position on the circuit board, Photo **B**. Pick and place machines can accurately position thousands of components per hour. They are much faster and more accurate than human workers.

B *A pick and place machine putting components on a PCB*

Making enclosures

In the school workshop, a laser cutter may be used to cut out the parts needed to make an enclosure. In industry, CNC machines used during the manufacture of enclosures may include injection moulding machines (see topic 7.6), lathes, drills and integrated machining centres, which can carry out both milling and drilling.

What is CIM?

The machines used for CAM need to be programmed. If a product has been designed using CAD, computer software can analyse the drawing and create the control program. This can save hours of manual programming time. This linking of CAD and CAM is called **computer integrated manufacture (CIM)**. CIM combines all the advantages of CAD and CAM.

CIM can be used to reduce the amount of time needed to develop a product and bring it to market. For example, CAD drawings can be sent directly to machines on the shop floor as soon as they are finished.

Summary

CAM is the use of computers to operate CNC machine tools.

CNC machines can be faster, more consistent and more reliable than manual machines. However, they need time to program and are more expensive.

CNC machines include pick and place machines to put components on circuit boards and routers to mill circuit boards for one-off products.

CIM is the use of CAD to design a part followed by the use of CAM to manufacture it on CNC machines.

Remember

Development of CNC equipment

1940 First computer

1952 Prototype of the first CNC machine

1954 Programming code developed for CNC

1957 First CNC machine tools become commercially available

Activity

Create a cartoon strip that shows the sequence of tasks to be carried out when using CIM to design and make the sides for an enclosure using a CAM router. This should be suitable for use to explain CIM to year 7 pupils.

AQA Examiner's tip

You should be able to list three advantages explaining why CAM is used.

kerboodle!

7.5 Manufacturing processes 1

For enclosures made out of thermoplastic materials, such as acrylic, there are several possible manufacturing processes. Process selection will depend on a number of factors, such as cost, the quantity to be made and the design of the enclosure.

Objectives

- Describe the rapid prototyping processes.
- Describe how laser cutting can be used to make enclosures.
- Describe how to carry out line bending using a sheet of plastic.

Rapid prototyping

Rapid prototyping is the manufacture of a working model of a part directly from a CAD program using a computer-controlled machine. It can produce a complicated three-dimensional shape from the CAD model. These parts are normally used to evaluate the design, but they can be used for enclosures. There is a range of different processes that can be used to make the part, including selective laser sintering (SLS), three-dimensional printing and stereolithography (SLA). These processes often take many hours to produce a part. They are only suitable for making one-off products where cost is not a limiting factor.

A *Rapid prototyping in school*

Selective laser sintering

In this process a laser is used to melt precise areas in a layer of powdered material. These harden on top of previous layers, fusing more and more layers together until the component is complete. The individual layers are incredibly thin. The final solid product may contain millions of layers. The process is driven by a CAD file and can produce very accurate finished products without the need for elaborate tooling and with very little waste.

Three-dimensional printing

This process uses a technique that prints layer upon layer of a material until it builds up a three-dimensional solid. Prototypes made from wax, starch or some thermoplastics can be printed in complex shapes restricted only by what can be generated on the computer.

Key terms

Rapid prototyping: making an example of a product for evaluation, normally on a single computer-controlled machine.

Laser cutting: using a laser to cut out a shape by melting or vaporising the material along the cut line.

Bending: forming an angle or curve in a single piece of material.

Stereolithography

This method of rapid prototyping uses a laser to scan a bath of photosensitive resin. As the laser scans across the surface of the liquid, it solidifies selective areas and then moves to the next layer. A three-dimensional model is built up below the surface of the liquid and is subsequently removed.

Laser cutting

Laser cutting is a relatively new process which is now being used in all levels of production. Two-dimensional CAD images can be cut out from many materials using a computer-controlled laser beam which melts through the material, Diagram **B**. The strength of the beam and the speed at which it moves determine the depth of cut. As no contact is made with the material being cut, no physical clamping is required. Very accurate two-dimensional products can be produced by this process. This can be used to make the sides of an enclosure, but these will then need to be joined together to make the finished enclosure. As this assembly can require a lot of effort, it is only used for one-off or small batch production.

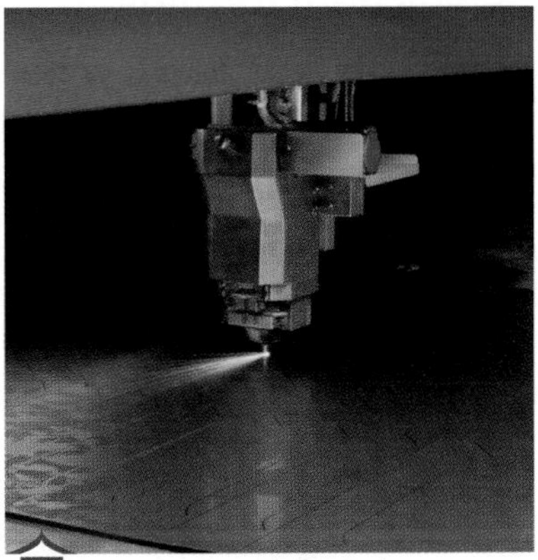

B *Laser cutting of plastic sheet*

Line bending

Line **bending** involves heating a sheet of thermoplastic material along a line, using a heating element. As the plastic heats, it softens, allowing it to be bent. As it cools, it will retain its shape. Line bending is often carried out in schools using acrylic sheet, because it is a low-cost process.

It is a good idea to make a wooden former to ensure accurate bending, Diagram **C**. It is important to allow the plastic to cool slightly before it is removed from the former – otherwise, if it is still hot it may sag, changing its shape.

Strip bending can be carried out by hand for one-off or small batch production. A simple enclosure can be made in two U-shaped parts, which can be attached to one another using a suitable adhesive.

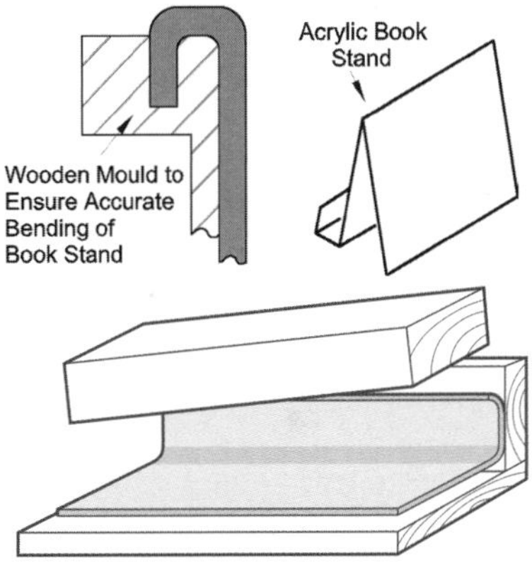

C *Line bending mould*

Activity

Using the internet, find videos of the different rapid prototyping processes in use. Compare the shapes that can be produced by the different processes and the materials used.

Summary

Rapid prototyping is used to make one-off parts in complex three-dimensional shapes from CAD drawings. The processes used include selective laser sintering, three-dimensional printing and stereolithography.

Laser cutting can be used to cut out simple two-dimensional parts that can be assembled to make an enclosure.

Line bending is a quick and easy process to form plastic sheet using heat. It can be used to make simple bends, reducing the number of parts needed to make an enclosure.

AQA Examiner's tip

You need to be able to describe the rapid prototyping processes, even if you haven't used them.

kerboodle!

7.6 Manufacturing processes 2

Two further processes that can be used to manufacture the enclosures for electronic products are **vacuum forming** and **injection moulding**.

Vacuum forming

Vacuum forming is used to make many different products, including enclosures for electronic products, packaging, helmets and baths. It uses sheets of thermoplastic. These are heated to make them flexible, formed over a **mould** using a vacuum, and then cooled to become hard again, Diagram **A**.

The moulds used are often made from wood or MDF and are much cheaper than those used for injection moulding. The sides of the mould must slope to allow the plastic product to be lifted off or pulled out, Diagram **B**. This slope is called the draft angle; it should be between 5 and 10°. If there were no angle, the plastic product might stick to the mould. The corners of the mould should have a radius. Any recesses must have small vent holes drilled in them to prevent trapped air from stopping the plastic sheet forming.

Vacuum forming can only be used to make shapes of simple profiles, as any overlaps would cause the plastic to be stuck to the mould. Unlike injection moulding, it cannot make reinforcing fins inside the shape.

Vacuum forming is commonly used for one-off and batch production. It is not often used for mass production as the shape has to be cut out from the plastic sheet. This adds an extra production step, and therefore extra cost, compared to processes that make the part using liquid plastic.

Objectives

- Describe the vacuum forming process.
- Describe the injection moulding process.
- Explain the differences between vacuum forming and injection moulding.

Key terms

Vacuum forming: forming a thermoplastic sheet over a mould, using heat and a vacuum.

Injection moulding: the process of making plastic parts by forcing liquid plastic into a mould and allowing it to solidify.

Mould: a former used to shape a part.

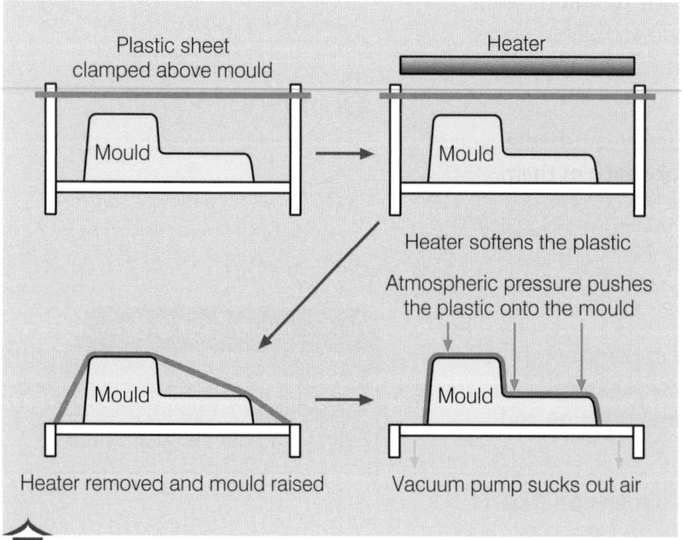

A *Vacuum forming*

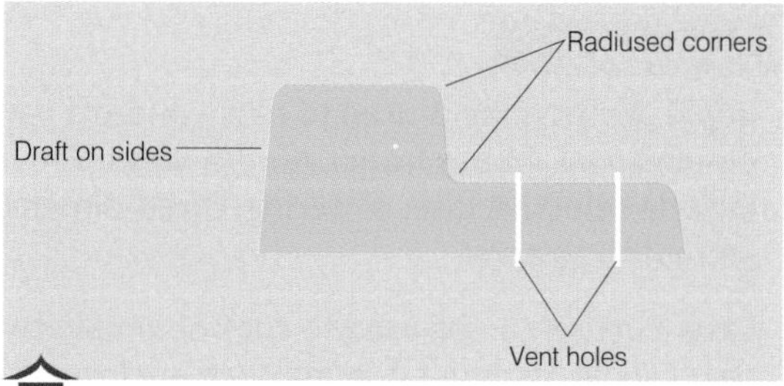

B *Vacuum forming mould*

Injection moulding

The injection moulding of thermoplastics is very versatile, producing items such as bowls, cases for remote controls, model construction kits and enclosures for games consoles.

Diagram **C** shows how injection moulding is carried out. Plastic granules are loaded into a feed hopper. From here they drop into the barrel, where the rotating screw thread pushes the plastic along. It is melted by the heaters. At the end of the barrel the cone compresses the plastic and it is injected into the mould. The mould is cooled and the component is ejected. The mould can be moved back into position to make the next part within seconds.

Although the equipment and moulds are very expensive, the process is very fast, and complex shapes can be made. This means that normally it is only used when making thousands or tens of thousands of parts, so that the high cost can be divided between all of the parts made. However, it is sometimes used for making small quantities of expensive parts for special applications, such as a spacecraft.

It is normally easy to identify an injection moulded part, as the sprue point where the plastic was injected is often visible. There may also be a split line visible if the sides of the mould did not fit together perfectly.

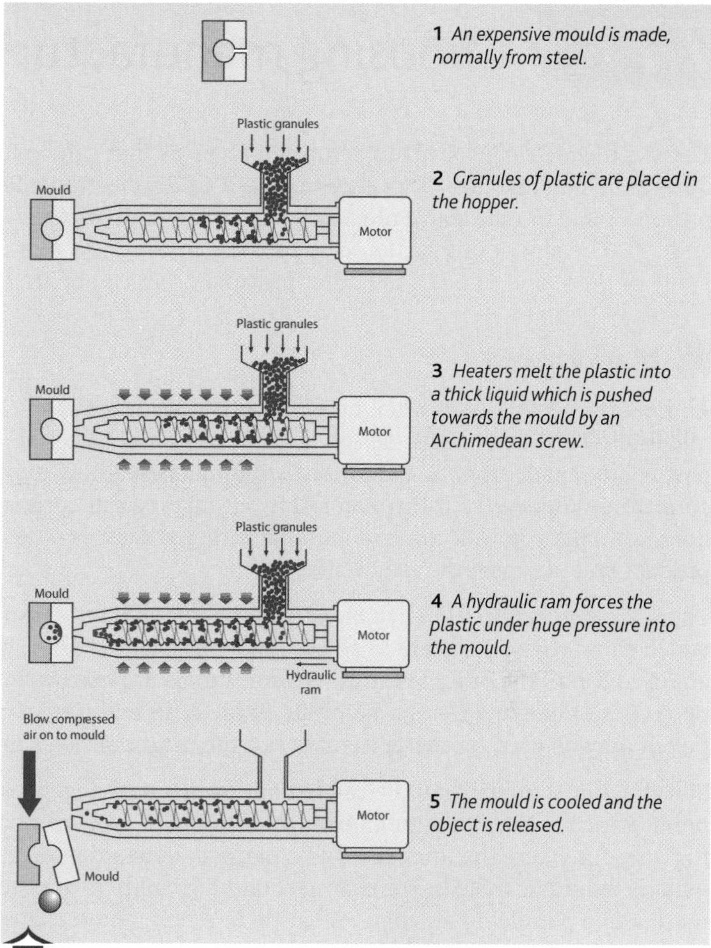
1 An expensive mould is made, normally from steel.
Plastic granules
Mould
Motor
2 Granules of plastic are placed in the hopper.
Plastic granules
Mould
Motor
3 Heaters melt the plastic into a thick liquid which is pushed towards the mould by an Archimedean screw.
Plastic granules
Mould
Motor
Hydraulic ram
4 A hydraulic ram forces the plastic under huge pressure into the mould.
Blow compressed air on to mould
Motor
Mould
5 The mould is cooled and the object is released.

C *The injection moulding process*

D *The case of a remote control is an example of a product made by injection moulding*

Activities

1. Find three examples of vacuum-formed products. Identify the type of thermoplastic that may have been used to make each of them.
2. Find a plastic product that has been made by injection moulding. See whether you can identify the sprue point.

Summary

Vacuum forming can be used to make simple three-dimensional shapes. It uses a cheaper mould than injection moulding but it is a slower process and normally requires more labour.

Injection moulding involves forcing liquid plastic into a mould. It can be used to make complicated, three-dimensional shapes.

The equipment and mould costs for injection moulding are very high, so it is normally only used either when making large numbers of parts or for high-cost parts for special applications.

AQA Examiner's tip

You will need to be able to explain the differences between injection moulding and vacuum forming in making plastic products.

kerboodle!

7.7 Choosing manufacturing processes

There is a wide range of manufacturing processes that can be used during the manufacture of an enclosure in a GCSE Electronic Products project. Some of these have already been explained in topics 7.4, 7.5 and 7.6. The aim of this section is to provide an overview that includes many of these and some of the other processes that can be used.

Objectives

- Explain why marking out is important.
- List a range of tools that may be used for marking out.
- Select a suitable manufacturing process for the manufacture of an enclosure.

Key term

Marking out: drawing the design of a part onto the materials that it will be made from.

Marking out

An important step that is often overlooked is **marking out** the materials. This is where the design is drawn onto the materials that the part will be made from. It cannot be over-emphasised how important correct marking out is. If the material is not marked out correctly, it does not matter how accurate a manufacturing process is – the final product will not meet the requirements.

There are two types of line which need to be marked out: cutting lines, which show where parts are to be cut; and construction lines, which are used for all the other features. Different tools are used to make the two types of line in different materials, Table **A**. In addition, a centre punch may be used for metal to mark out the centre of holes in metals.

Marking out should always be carried out from a true edge or specified point, which is called a datum. A range of different tools can be used to work out where the lines should be marked, depending on exactly what is required, Table **B**. For example, odd leg callipers are used to mark a line parallel to an edge on plastic or metal, and tri-squares and engineer's squares are used to mark lines at right-angles to an edge.

A *Tools for marking out*

Tool to use	Cutting line	Construction line
Wood	Marking knife	Pencil
Plastic	Scriber	Felt tip or wax crayon
Metal	Dot punch, making dots at 5 mm intervals	Scriber

Activities

1. Mark out a template for a steel panel on a piece of card. It should be 100 mm square with a radius of 5 mm at each corner. It should have four holes, one in each corner, each of diameter 10 mm, each located 15 mm from the two edges.
2. Identify the different processes that would be needed to manufacture the following products:
   - the damage-proof enclosure for an alarm, made from low a carbon steel sheet.
   - the body of a kettle, made from a thermoplastic.

B *Tools used for marking out*

Tool	Wood	Plastic	Metal
Angle plate		Yes	Yes
Centre punch			Yes
Compass	Yes	Yes	
Dot punch			Yes
Engineer's rule	Yes	Yes	Yes
Engineer's square		Yes	Yes
Marking gauge	Yes	Yes	
Mitre square	Yes	Yes	
Odd leg callipers		Yes	Yes
Sliding bevel	Yes	Yes	Yes
Surface plate		Yes	Yes
Templates	Yes	Yes	Yes
Tri-square	Yes	Yes	
Vee block		Yes	Yes

Choosing manufacturing processes

There are a large number of different processes and equipment. Many of these are suitable only for certain types of material or forms of material. A chooser chart, like that shown in Table **C**, can be used as a starting point in selecting the processes that could be used to carry out the required tasks. However, before making the final choice of what tool to use, it is necessary to consider the number of parts to be made, the accuracy needed and what equipment is available.

AQA Examiner's tip

Measure twice, cut once – after you have finished marking out a part, check every measurement again to make sure that they are all correct. This is a lot quicker than having to change a part if it is wrong.

C *Methods used to manufacture different materials in a school workshop*

Type of process	Material	Pine (softwood)	Mild steel sheet	Mild steel bar	Aluminium alloy	Melamine	Acrylic sheet	Polypropylene	Alumina	Fibreglass (GRP)
	Tool or Machine / Material type	Wood	Ferrous material	Ferrous material	Non-ferrous material	Thermoset	Thermoplastic	Thermoplastic	Ceramic	Composite
Material removal	File	No	Yes	Yes	Yes	Maybe	Yes	Maybe	No	Maybe
	Drill	Yes	Yes	Yes	Yes	Yes	Yes	Yes	No	Yes
	Engineer's lathe	No	No	Yes	Yes	No	Yes	No	No	No
	Milling machine	No	Yes	Yes	Yes	No	Yes	No	No	No
Cutting	Coping saw	Yes	No	No	No	No	Yes	Yes	No	No
	Hacksaw	No	Yes	Yes	Yes	Yes	Yes	Yes	No	Maybe
	Guillotine/shears	No	Yes	Maybe	Yes	Maybe	No	No	No	No
	Laser cutting	No	No	No	No	No	Yes	Yes	No	No
Shaping and forming	Sand casting	No	N/A	N/A	Maybe	No	No	No	No	No
	Injection moulding	No	No	No	No	No	Yes	Yes	No	No
	Moulding	No	No	No	No	No	No	No	No	Yes
	Sintering	No	No	No	No	No	No	No	Yes	No
	Hammer	No	Yes	Yes	Yes	No	No	No	No	No
	Press	No	Yes	Yes	Yes	No	No	No	No	No
	Strip bending	No	No	No	No	No	Yes	Yes	No	No
	Vacuum former	No	No	No	No	No	Yes	Yes	No	No
	Compression moulding	No	No	No	No	Yes	No	No	No	No
Joining	Welding and brazing	No	Yes	Yes	Yes	No	Maybe	Maybe	No	No
	Adhesives	Maybe	Maybe	Maybe	No	Maybe	Yes	Yes	No	Yes
	Riveting	No	Yes	Maybe	Yes	No	No	No	No	No
	Screwed fasteners	Yes	Yes	Yes	Yes	Yes	Yes	Yes	Yes	Yes

Key: Yes = can normally be used with this material; maybe = sometimes can be used with this material; no = not suitable for use with this material

Summary

Marking out is the most important activity towards achieving the correct dimensions of a product. There is a wide range of tools available to support the marking out of different materials and different types of dimension.

A chooser chart is a useful starting point for selecting the tools and equipment that could be used to make a product. There is a wide range of equipment available to carry out different manufacturing processes. Each tool has different characteristics. Some equipment is only suitable for a small range of materials.

7.8 Housing electronic circuits

In topics 7.4 to 7.7 we have covered a wide range of different manufacturing processes that may be used during the manufacture of an **enclosure**. The aim of this section is to demonstrate how a simple enclosure can be manufactured and put together.

This is based on a one-off enclosure being made to house the control system of a game where a player has three lives, indicated by LEDs lighting. It is made from coloured acrylic sheet. Any sheet material capable of being joined with an adhesive could be used, although other materials may need to be cut using different processes in step 1.

Constructing an enclosure

Step 1:
Cut out the parts needed to make the enclosure using a laser cutter.

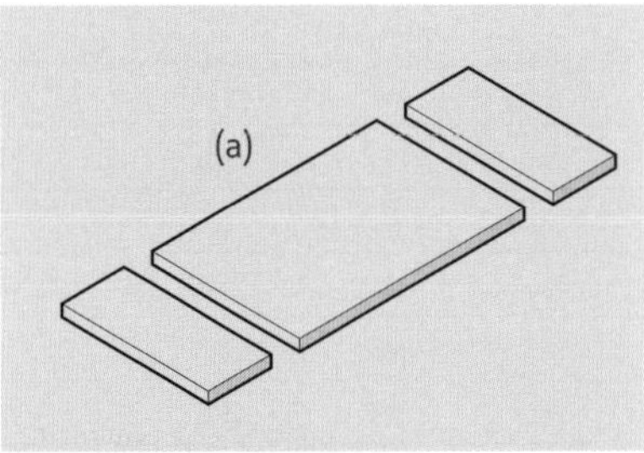

Step 2:
Using adhesive, attach the ends to the base, Figure **(a)** and **(b)**. Make sure they are at 90° to the base.

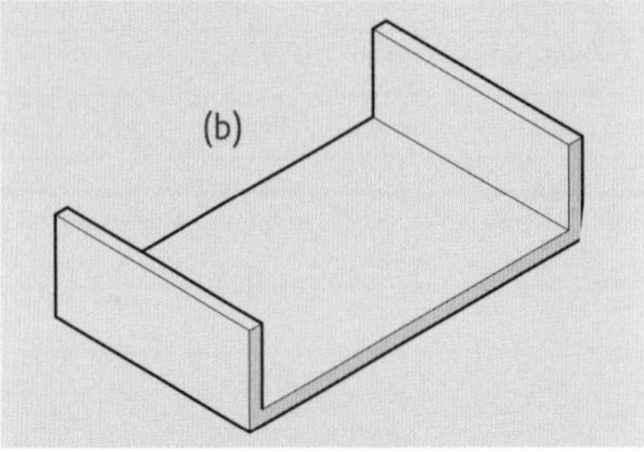

Step 3:
Using adhesive, attach the long sides to the base, Figure **(c)**. Any overlap should be carefully removed with a file.

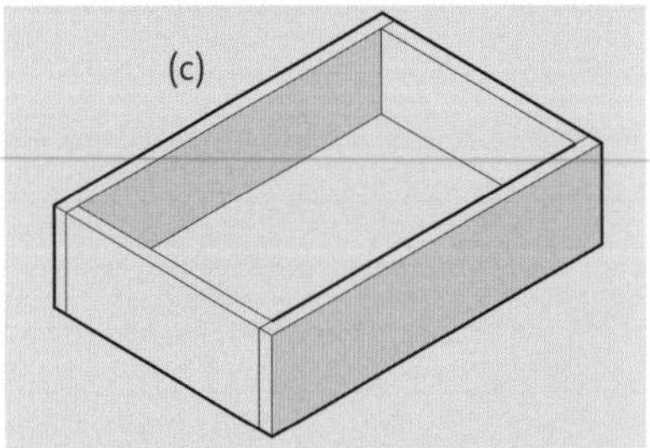

Step 4:
Add an insert to both of the long sides, Figure **(d)**. This will support the lid. The lid should be the same size as the base and should be a snug fit. If it is too large, carefully remove any excess material using a file.

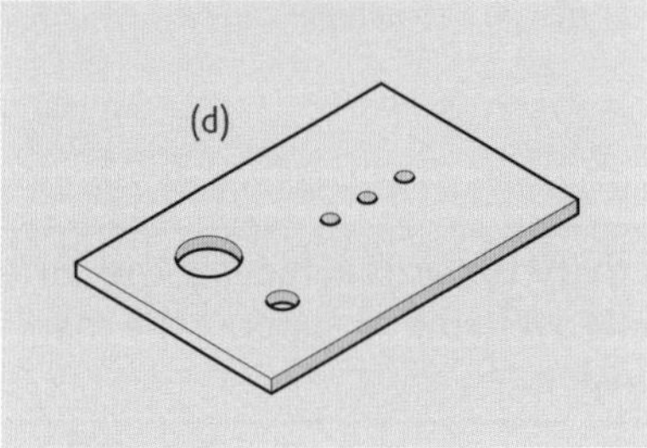

Objectives

- Describe how an enclosure can be built.
- Describe how to secure the PCB.
- Describe how to secure the battery.

Key terms

Enclosure: a container for an electronic device. This includes cases, graphic displays, garments and soft containers.

Step 5: :
Using an engineering ruler, scriber and tri-square, mark out the positions for the holes needed in the lid. Drill the holes in the lid using a pillar drill, Figure **(e)**. If there are any rough edges round the holes, smooth these using a round file.

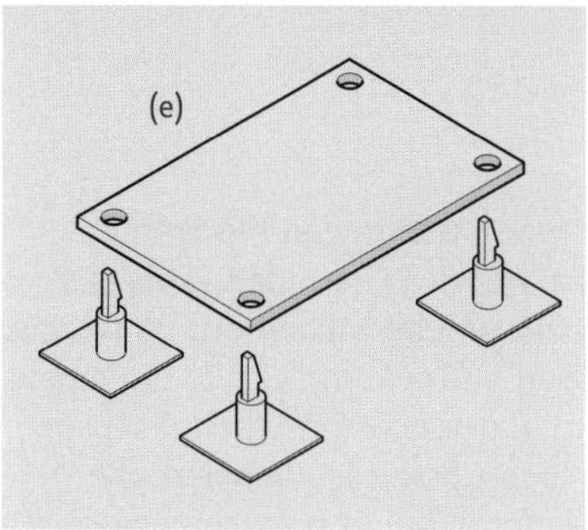

Step 6:
Secure the PCB in the enclosure. This will stop the circuit being damaged if the product is moved. A simple method is to use a PCB pillar with self-adhesive feet, Figure **(f)**. These stick through holes in the PCB to hold it in place. The pillars are inserted into the PCB before being stuck down.

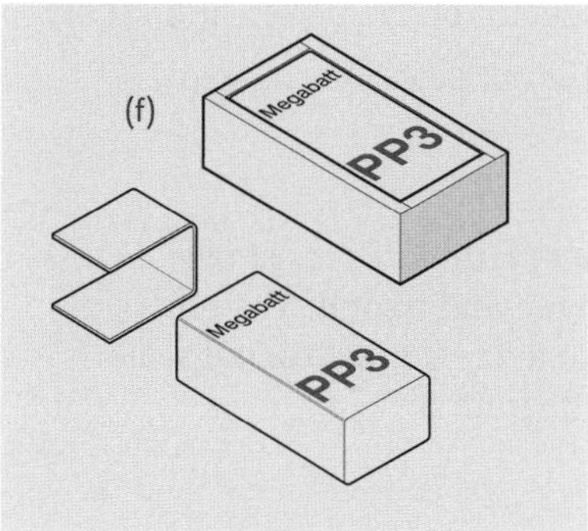

Step 7:
Secure the battery in the enclosure. This will stop it rattling or damaging the circuit when the product is moved. However, it has to allow the battery to be changed when the charge runs out. Figure **(g)** shows two different options that could be used to secure a PP3 battery.

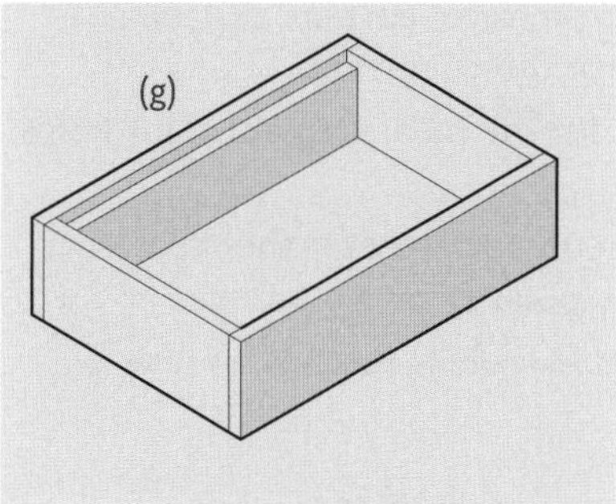

Step 8:
Attach the components to the lid and close the enclosure, Figure **(h)**. Make sure that any leads are tied together using plastic ties or clips, so that they are neat and tidy and cannot get tangled up and damaged.

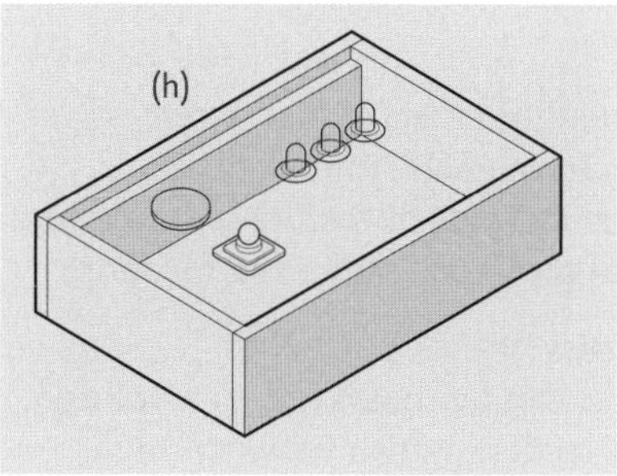

Summary

An enclosure can be manufactured as a group of parts that need to be assembled together.

A PCB can be secured in an enclosure using PCB pillars.

A battery can be secured inside an enclosure by building a battery compartment or making a battery clip.

Activity

Using the internet or, if available, samples of components, identify the different methods that can be used to attach a range of input and output components to an enclosure.

AQA *Examiner's tip*

You should know how to fit switches, other inputs and outputs to an enclosure, and should know the tools to be used when making an enclosure.

kerboodle!

7.9 Testing

One of the final tasks to be carried out when making an electronic product is to make sure that the product works as intended. This is a normal part of quality control. Ideally, this can be carried out by testing the product in the way that it will be used – for example, playing music from an MP3 player or checking that an alarm goes off when a sensor is activated. Where this type of testing is not possible, or where the circuit does not appear to be working as intended, it may be necessary to use a **multimeter** to check that the circuit or the parts within it are working correctly.

Objectives

Use a multimeter.

Explain how a multimeter can be used to identify possible causes of failure in a circuit.

Multimeters

Multimeters are used to measure voltage, current or resistance, Diagram **A**. There are two types: analogue and digital. The analogue scale shows the value by the position of a needle; the digital scale simply shows the value as a number, Diagram **B**.

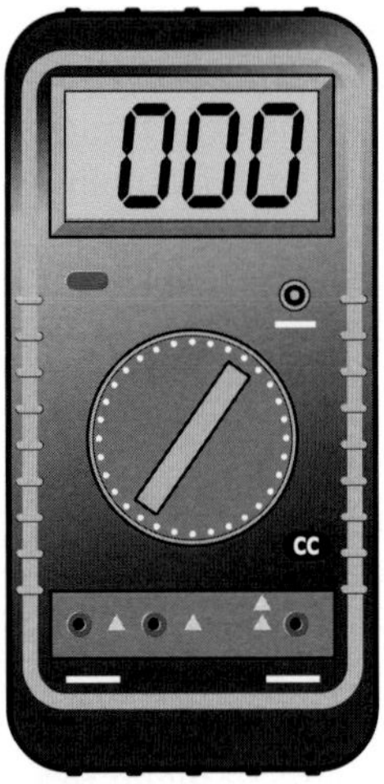

A *A multimeter*

How to use a multimeter

1 Decide on the type of measurement (voltage, current, resistance) and check whether it is alternating or direct current.
2 Choose the highest range available for the measurement. This helps to prevent damage to the equipment.
3 Connect the multimeter into the circuit correctly for the type of measurement, taking account of the polarity of the probes.
4 Adjust the range down until a suitable scale is reached to take a reading.

Examples of what a multimeter can be used for

Testing systems blocks

One of the most effective methods of testing a circuit that is not working correctly is to test the outputs from each systems block contained within it. This can reduce the amount of testing required and narrow down the areas where potential problems may exist for more detailed investigation.

The signal level between systems blocks can be measured as a voltage. For digital outputs, it is only necessary to distinguish between high and low. If a multimeter is not available, this can be carried out with a **logic probe**. For signals that are changing more than about once per second – for example, the output of an astable circuit – these cannot be read accurately with a multimeter and it may be necessary to use an **oscilloscope**.

Key terms

Multimeter: a device that can be used to measure current, voltage or resistance.

Logic probe: a device that can be used to determine whether a digital signal is high or low.

Oscilloscope: a piece of equipment with a visual display that can be used to accurately measure rapidly changing signals.

Ohmmeter: a device used to measure resistance.

Testing for short circuits and open circuits: continuity testing

In a short circuit, current will be flowing between two points where it should not be. In an open circuit, current will not be able to flow between two points where it should be able to flow.

Continuity testing uses a multimeter to measure resistance to show whether a current can flow between two points, that is, as an **ohmmeter**. By testing at opposite ends of a PCB track, if there are no breaks, the value measured will be zero ohms. If there are breaks in the track, the reading will show a high resistance. This can also be used to detect where short circuits exist between two points that should not be connected, to check wiring for breaks or to check the correct operation of switches within the circuit.

Component testing

Resistance can be measured to check that diodes are positioned with the correct polarity. They have low resistance in the direction that they allow current to flow and very high resistance in the other direction.

Resistance can be measured to check that sensors such as thermistors or light-dependent resistors are changing.

The correct function of other components, such as transistors or thyristors, can be checked by measuring current or voltage, as suitable for the individual component.

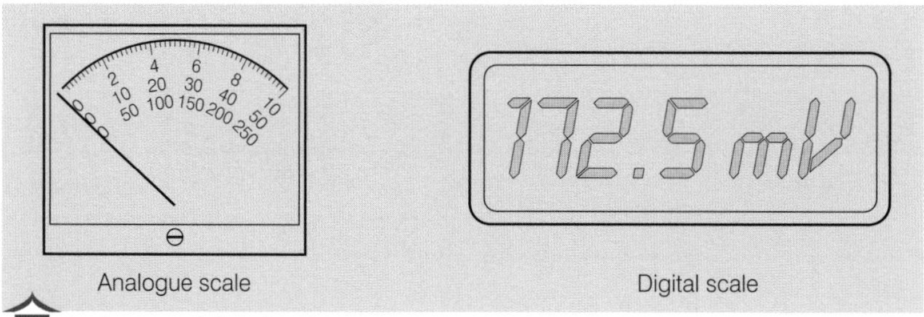

B *Analogue and digital scales*

Remember

Accuracy of analogue readings

An analogue scale is non-linear. The most accurate readings are for low values shown on the scale.

Activity

Build one of the circuits shown in chapter 3 (not an astable circuit). Using a multimeter, measure the current in the circuit, the voltage across the different components and the resistance between different points in the circuit.

AQA Examiner's tip

When assembling a circuit, it is a good idea to test each systems block as it is assembled on the PCB. This identifies any problems very quickly and allows them to be repaired.

Summary

The voltage output from systems blocks within a circuit can be measured to check that they are working correctly.

A multimeter can be used as an ohmmeter to carry out continuity testing to detect short circuits and open circuits.

Multimeters can also be used to check that diodes are positioned with the correct polarity or to check the values of individual components.

AQA Examination-style questions

AQA Examiner's tip Avoid one-word answers unless there is only one mark for the question. To gain full marks you need to give reasons for your answer.

1. Describe two quality checks that could be made on a completed circuit board.

(2 × 3 marks)
(Total for Q1: 6 marks)

2. Shown below are areas of electronic design where computers could be used. Choose **three** areas from the list, stating when each could be used and explain **one** advantage for each choice.

Circuit simulation **PCB design** **Case design** **CAM**

(a) ....

When used ....

(1 mark)

Advantage ....

....

(2 marks)

(b) ....

When used ....

(1 mark)

Advantage ....

....

(2 marks)

(c) ....

When used ....

(1 mark)

Advantage ....

....

(2 marks)
(Total for Q2: 9 marks)

3. This question is about the social, moral and environmental aspects of mobile telephones.

You are advised to spend about 10 minutes on this question.

Mobile telephones have changed the way we communicate at work and during our leisure time. Compare the advantages and disadvantages of the increased use of this type of technology. *(10 marks)*

4. During recent years the use of ICT and electronic control systems has revolutionised manufacturing.

Explain **one** advantage and **one** disadvantage that these developments have had for the environment.

(2 × 3 marks)
(Total for Q4: 6 marks)

5. This question is about the development of electronic products.

The development of electronic products is having a major impact on society, an example being personal music systems.

A

(a) Describe the advantages **and** disadvantages the development of electronic products has had for teenagers.

(4 marks)

(b) Most of these electronic products use batteries. Why is it important to dispose of batteries correctly to protect the environment?

(4 marks)
(Total for Q5: 8 marks)

6. Describe the advantages of using CAD when designing printed circuit boards. *(4 marks)*

7. Electronic games have changed the way we spend our leisure time.

(a) Describe how electronic games have changed in recent years. *(3 marks)*

(b) Explain the impact electronic games are said to have had on personal health and on social and family life.

(i) Personal health *(2 marks)*

(ii) Social and family life *(2 marks)*

(c) The rapid rate of change of electronic products means that some products soon become obsolete.

(i) What advantages does this have for the manufacturer? *(3 marks)*

(ii) What disadvantages does this have for the manufacturer? *(3 marks)*

(Total for Q7: 13 marks)

8. A manufacturer wants to go into the commercial production of an electronic timer for use when cooking.

He is considering either vacuum forming or injection moulding the case.

Compare the advantages **and** disadvantages of the two methods, and state the factors the manufacturer might need to consider.

(10 marks)

9. Health and safety is important when making a printed circuit board.

Identify **two** different hazards to be addressed when soldering the components in place.

Explain the precautions that need to be taken.

(6 marks)

10. Circuits are often trialled using temporary construction methods such as a prototyping board, see Diagram **B**.

B

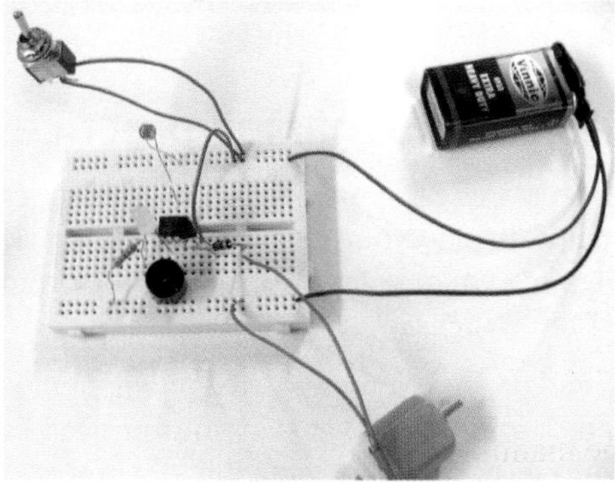

An alternative is to simulate the circuit on a computer.
Give two advantages of each method of trialling a circuit.

	Prototype board	Computer simulation
Advantage 1		
Advantage 2		

(8 marks)

11. Describe the main stages of making a circuit board using a method with which you are familiar. At each identified stage, describe and evaluate any quality control and health and safety issues that might occur.

(10 marks)

UNIT TWO Design and making practice

What will you study in this section?

- understand how Controlled Assessment is organised and assessed
- list the steps of the design process
- explain why the design process is often not a simple sequential activity in practice.

What is design and making practice?

Design and making practice is the production of an electronic product which will allow you to display your design and making skills. It is worth 60% of the marks available for your GCSE, the other 40% being awarded for the written paper.

The exam board provide a set of contexts, from which you must choose a task. This task will be the focus of your project, so you need to choose one that interests you. The latest list of contexts and tasks is displayed on the AQA website (**www.aqa.org.uk**), in the Design and Technology: Electronic Products section.

You will need to complete your project within a time allocation of 45 hours.

Your design folder should consist of approximately 20 sides of A3 paper, or about 30 sides of A4 paper. In line with the marks available, you should allocate the time equally between the design folder and the making activity.

Your project will be assessed using the Controlled Assessment Criteria, which are listed in the text box. These are explained in the AQA Design and Technology Electronic Products specification. The first four assessment criteria will be explained in this chapter. Assessment Criterion 5, communication skills, will be assessed throughout your folder. Your teacher will be looking for good written and graphical communication, good use of technical language, the use of appropriate, subject-relevant software, digital images and your knowledge and use of drawing conventions.

Remember

The five Assessment Criteria

1. Investigating the design context
2. Development of design proposals
3. Making
4. Testing and evaluation
5. Communication

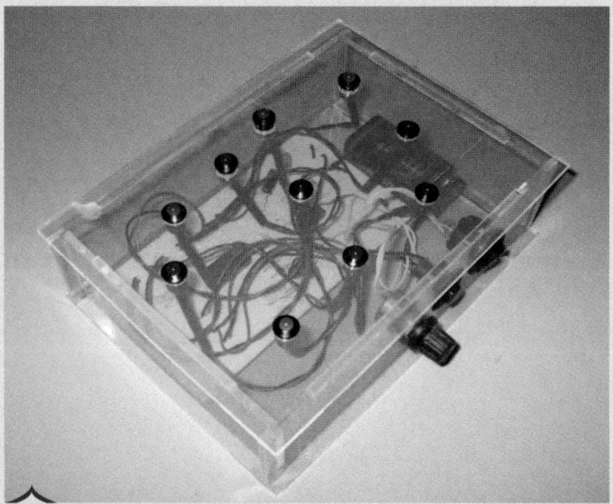

A *Example of a student project*

The design process

Your folder should follow the design process. This is a series of activities that need to be carried out to make sure that the product does what it needs to do. The individual activities will be explained in this chapter.

The design process in practice

In practice, the design process is rarely carried out as a simple sequence of tasks. Evaluation is carried out at every stage. Often things are found out as you go through the process that change what was needed at an earlier step. For example:

- The design brief might have to be changed following feedback from potential users during the research.
- The ideas developed might need to be changed when making the product, as it might not be possible to make the design to the accuracy needed.

These might mean that you have to jump back to earlier steps several times during the development of your product. It is a good idea to record the progress of the project as you go along, as it can be difficult to remember all the details and the reasons for your design decisions four weeks later!

Some effective ways of recording your progress include sketching as much as you can; you do not have to be a great artist, you just need to communicate clearly. You could write notes. These should be short and sharp – use bullet points if you can. You could also take digital pictures using a camera or mobile phone. Pictures of your computer models, breadboard modelling or the soldering of your circuit add real value to your design folder.

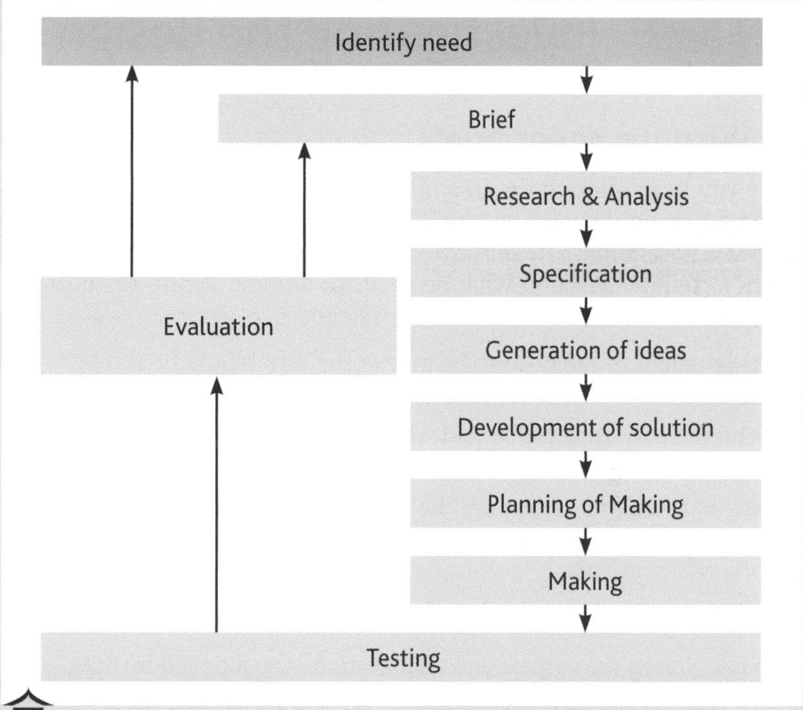

B *The design process*

AQA Examiner's tip

Don't skip stages of the design process or jump directly to a solution; this will probably mean that your solution will not be as good as it could be.

Record your thinking, why you made decisions at each stage of the process in case you need to make changes later.

Activity

A company have decided to develop a new mobile phone. For each step of the design process shown above, identify at least one thing that could happen, or one result or finding, that might mean that they would need to go back to an earlier step in the design process.

8 Design and making practice

8.1 Investigating the design opportunity

What is a design brief?

Once you have selected the task that you will carry out, you need to prepare a **design brief**. This can be a short statement, often no more than one paragraph. The purpose of the design brief is for the user or **client** to tell the designer what he needs to design. Ideally, the design brief should include:

- a statement of the need to be met or the problem to be solved – that is, the **function** that the product must carry out
- who the end user will be and who would buy the product, if these two are different – the target market
- any **constraints** – these are the things that limit what you can make
- the important product features or **user needs**.

Examples of a design brief

Tom has chosen the context and task from the list supplied by AQA which relates to a counting activity. The school wants to count how many students use the library at lunchtime, so that they can have some data to measure its popularity and to help them decide how to develop the facility. He must design and make a system which will count and display the numbers of students entering the library at lunchtime. The project could be extended by the device keeping a running total of pupil numbers in the library at any one time. Tom's brief is shown in Figure A.

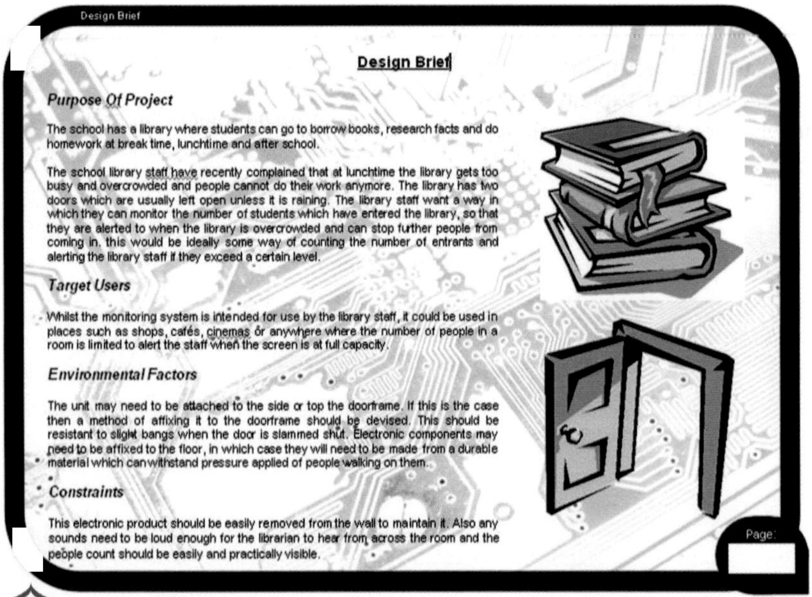

Design Brief

Design Brief

Purpose Of Project

The school has a library where students can go to borrow books, research facts and do homework at break time, lunchtime and after school.

The school library staff have recently complained that at lunchtime the library gets too busy and overcrowded and people cannot do their work anymore. The library has two doors which are usually left open unless it is raining. The library staff want a way in which they can monitor the number of students which have entered the library, so that they are alerted to when the library is overcrowded and can stop further people from coming in. this would be ideally some way of counting the number of entrants and alerting the library staff if they exceed a certain level.

Target Users

Whilst the monitoring system is intended for use by the library staff, it could be used in places such as shops, cafés, cinemas or anywhere where the number of people in a room is limited to alert the staff when the screen is at full capacity.

Environmental Factors

The unit may need to be attached to the side or top the doorframe. If this is the case then a method of affixing it to the doorframe should be devised. This should be resistant to slight bangs when the door is slammed shut. Electronic components may need to be affixed to the floor, in which case they will need to be made from a durable material which can withstand pressure applied of people walking on them.

Constraints

This electronic product should be easily removed from the wall to maintain it. Also any sounds need to be loud enough for the librarian to hear from across the room and the people count should be easily and practically visible.

Page:

A *A design brief for a library system by Tom*

Objectives

Explain what a design brief is and the information it should include.

Select a design brief which will allow them to display a wide range of skills and address all the assessment criteria.

Explain how to analyse a design brief and identify other design needs.

Key terms

Design brief: a short statement of what is required.

Client: the person that the work is being carried out for.

Function: what the product is intended to do.

Constraints: things that limit what you can make.

User needs: the things that the customers require the product to do.

AQA Examiner's commentary

Tom's design brief identifies all the key points – the function, user/target market, constraints and user needs. It is sufficiently broad that he will have lots of design options. This is a very good start to the project.

Analysis of the design brief

The next task is to analyse the brief. This is an opportunity to examine the design need in detail and to consider a really wide range of factors that might affect your design thinking. One way to do this is to create a word web or spider diagram, Figure **B**. In this, you ask as many questions as possible and carry out some creative, 'blue sky' thinking. This can help you to come up with a really original idea later in the design process.

B *A mind map*

Activity

Choose two electrical products at home, such as an electric toothbrush and an egg timer. Write separate design briefs for them. Explain the differences between these two briefs and what you think these might mean for the design of the product.

Summary

The design brief is a short statement of the need that must be met.

The design brief should include features that are important to the design, such as constraints and user needs.

We analyse the design brief to start creating a list of needs that the product must meet. The analysis may also create a list of other questions that need to be answered to enable the design of an effective product.

AQA Examiner's tip

When deciding what project to carry out:

- choose a project that really interests you
- make sure that you have sufficient electronics knowledge to enable you to make it
- ask your teacher if it is a sensible choice – you will be relying on your teacher for help and advice
- do not rely on other students, friends and relations to advise you – it's your work.

8.2 Research and analysis

While analysing the brief, you will probably have started to identify a lot of questions that you need to answer to design a successful product. For a simple product, there may be just 20–30 questions. For a complex product with many components, such as a car, there may be thousands of questions to answer. One way of checking that you have identified questions for every type of need is to use the acronym **ACCESS FM**, as shown in Table **A**. Alternatively, you could check that you have answered the **5 Ws & H**, shown in the Remember box.

A *Questions that might be asked to identify possible design needs: ACCESS FM*

Type of need	Example of a question that may be asked
Aesthetic	What colour is the product? What finish or texture does the container have?
Cost	How much should the product cost?
Customer	Who is the product designed for? When, where and how will they use it?
Environment	Can the product be recycled? What power source does it use?
Safety	What safety standards does the product need to meet? What must be done to make sure that it is safe for the user?
Size	How big should the product be?
Function	What is it intended to do? How could it do this?
Maintenance	Do you have to be able to change batteries or take it apart to repair it?
Materials	What properties are needed from the materials that will be used to make the design?
Manufacturing	Are there certain processes that must be used to make the product?

The next step in the design process is to find the answers to these questions. We do this by carrying out research and analysing the findings.

Research

Research is the name given to the activities that you carry out to find the answers to the questions about the brief. Sometimes a single piece of research may answer more than one question, but other questions may need separate pieces of research. Research can be time-consuming and generate lots of paper, which will fill up your design folder quickly. You must be selective, making sure that it is relevant. You must also be concise. If you have a lot of research you do not have to include all of it in the main part of your folder. You could just analyse your findings and reference the sources in an appendix at the end of your design folder.

There are a lot of different types of research. In general, these fall into two categories: primary and secondary.

Objectives

Select research methods to investigate the brief.

Analyse research to enable you to write clear and specific design criteria.

Key terms

Research: activities investigating or clarifying the design needs.

Analysis: reviewing the research and deciding what it means for your product.

Remember

5 Ws & H

WHO will use it?

WHAT is it being used for?

WHY is it being used?

WHEN will they use it?

WHERE is it being used?

HOW is it being used? How is it made?

Primary research

This is where you identify the raw data or carry out measurements to reach your conclusions. For example, this could include:

- market research – identifying what similar products are available and their features
- product analysis – looking at or taking apart examples of products that address similar needs
- carrying out tests to determine how the product should function, such as breadboarding or modelling to see how components work
- carrying out questionnaires and interviews to determine user needs.

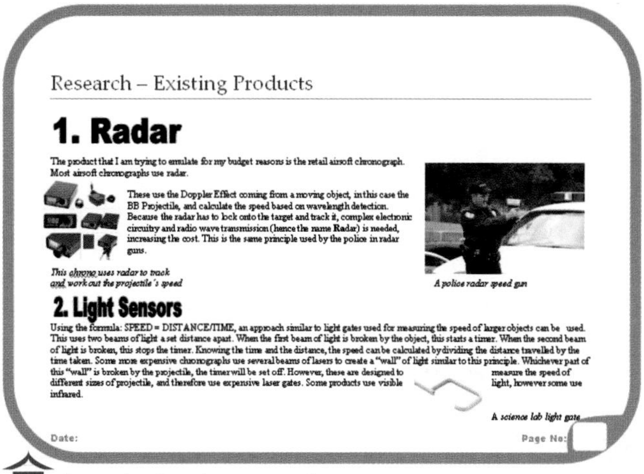

B *Example of student research*

Secondary research

This is where you use other people's sources of data. These can include textbooks, CDs or the internet. For example, it could include looking up the sizes that a product should be in a table of data.

If you use secondary research, you must make sure that you clearly indicate the source of the data otherwise you may be accused of plagiarism.

Your project will involve a form of electronic system and the container in which this is held, called an enclosure. Your research must cover both of these parts.

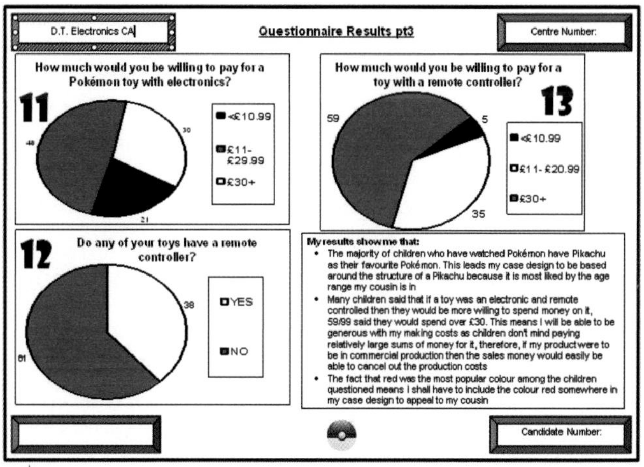

C *Example of student research*

Analysis

Analysis is not a separate activity – it is normally carried out at the same time as you find the results of your research activities. You should include comments on each piece of research showing what this means for the design, such as: 'As a result of this, the design should … '. This is because sometimes different types of need may contradict each other. If that happens you will need to know where each need has come from, so that you can decide which one is the more important.

Summary

Research should be selective, relevant and summarise your findings.

It can be carried out by primary or secondary methods.

You analyse research to be able to identify the most important factors for your initial specification.

Activity

Create a list of all the questions that you would need to answer if you were designing an electronic toothbrush.

kerboodle!

8.3 The design specification

What is a specification?

The specification is a list of all the needs that the product must meet. Most of these needs are answers to the questions that were identified when analysing the brief and investigated during the research. It should include, as a minimum, the design constraints and the wants of the user and market, relating to the **form** and function of the product. Table **A** is an example of a design specification for a keypad operated alarm system.

Ideally, you should normally include comments justifying why each of the needs is important. You could also indicate where that particular need was found during the research. There are two reasons for doing this:

- Some needs may have to be a compromise. For example, you might have to compromise between the size of the PCB, the number of batteries used and the size of the enclosure. You would either have to make a decision as to which piece of research was the more important, or find a compromise where the design need manages to satisfy all of the requirements.
- It can also be very important where, due to other issues later in the design process, it is necessary to go back and change a design.

Why is the specification important?

The specification sets the **design parameters** that you have to work to. This means that it limits what you can do. If any relevant needs have been missed from the specification, you may design a product that does not do what potential customers or the market need it to do.

Using the specification to evaluate the product

Design ideas and the final design proposal should be tested against the specification, to see whether they meet all of the needs. For this reason, it is important to make sure that every need is **quantifiable** and can be objectively tested. One way of doing this is to make sure that all of the listed needs have five characteristics, known by the acronym **SMART**:

- **Specific.** For example, for an alarm circuit, the designer would not say 'it needs to work well', as different people may have different understandings of what this means. The designer could say 'it must be activated when a door is opened', as all engineers would understand this.
- **Measurable.** For example, the designer would not say that the enclosure for an electronic product should be small, as different people will interpret this differently. The designer could say 'the enclosure should have maximum dimensions of 100 × 75 × 25 mm'.

Objectives

- Explain what a specification is and the information it includes.
- Explain what SMART stands for and how it should apply to a specification.
- Explain how the specification may be used to evaluate products.

Key terms

Form: the size and shape of the product.

Design parameters: the values for characteristics that the design has to satisfy.

Quantifiable: measurable.

- **Achievable.** A common mistake is to use an exact measurement, such as 'the remote control must work from 10 metres away'. This would mean that if the remote control didn't work from 9.9 or 10.1 metres away, but worked at 10 metres, the product would pass! You should state either a range, such as 'the remote control should work from 3 to 10 m away', or for other criteria a maximum or minimum value, such as 'the remote control should work up to 10 m away'.
- **Realistic.** There would be no point in writing down a need that could never be achieved!
- **Time-bound.** For example, there can be a big difference between two products where one needs to work for one minute and the other for one year. For example, in the case of an alarm this could specify the amount of time that a siren should sound for.

A *Example of a specification for a keypad operated alarm system*

No.	Need
1	The product is a keypad controlled alarm system. It requires a three-digit code number to be entered to turn off the alarm.
2	The product should be controlled by a picaxe-20M microcontroller.
3	The product should be linked to two contact sensors and two motion sensors, using multi-core wire.
4	It should include five LEDs, to show whether the alarm is turned on and which sensor has been activated.
5	The control box for the product should be powered by three 1.5 V AA batteries and include the keypad.
6	It should include a relay to operate a siren, which can be heard up to 30 metres away.
7	The siren should be housed in a separate enclosure and powered by four D batteries.
8	It should include an on-off switch and a download socket, so that it can be reprogrammed.
9	The enclosure for the control box should be made of white HIPS, with a maximum size of 150 × 100 × 25 mm.
10	The enclosure for the siren should be made of yellow HIPS, with a maximum size of 150 × 250 × 100 mm.
11	The PCB, siren and batteries must be secured within the enclosure so that they cannot move around.
12	It should be possible to open both enclosures to change the batteries. However, it must be possible to close them to prevent water entering.
13	There must be no sharp edges on the enclosures.
14	The product must be suitable for one-off production.
15	The total cost of the parts used must be less than £10.

Activities

1. Choose an electrical product at home. Write the specification that you think the designer used for that product.
2. Look at the example specification below. Identify the SMART characteristics of each of the stated needs. How do you think this specification could be improved?

AQA Examiner's tip

Use the design specification as a checklist for your ideas. Ensure specifications generate testable and quantifiable outcomes.

Summary

The design specification is a list of all of the important needs that the product must meet.

To ensure that the design proposal and final product can be tested against the specification, the needs should be specific, measurable, achievable, realistic and time-bound.

kerboodle!

8.4 Systems analysis

Objectives

- Describe what a system is.
- List the parts of a system.
- Carry out a systems analysis.

Systems diagrams

As explained in chapter 1, a system is a collection of parts that exists to perform a function. At minimum, it contains an input, a process and an output. Any product that is an assembly of parts and interacts with its environment is in some way a system.

A systems diagram is a representation of how a system will work. It breaks down the functions to be carried out into simple categories and lists the parts needed to carry out each category. This means that if you can create a systems diagram, you can produce an overview of the functional parts needed in the design of any product that is a system. Figure **A** is an example of a simple systems diagram for a timing system.

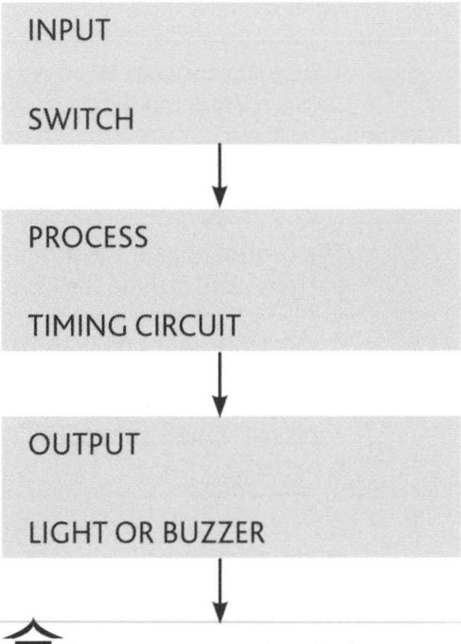

A *A systems diagram for a timing system*

Systems analysis

Systems analysis involves looking at the design requirements and identifying the possible inputs, processes and outputs that would be needed in an electronic product to satisfy these needs. It is a key activity in the design folder for your GCSE project and demonstrates your electronics knowledge. It should allow you to look at the different combinations that you could use to come up with an electronic solution. You might think that it is difficult to be creative with electronics, but this is a chance to come up with a combination that is really inventive.

If, for the product you are designing, you can think of three possible inputs, three possible processes and two possible outputs, you have a minimum of 18 possible combinations. Of these, you need to consider three or four combinations as ideas and choose one for further development. This is sometimes called Morphological Analysis.

Figure **B** shows a simple systems analysis prepared by a student called Jaymil, for the timing system shown in Figure **A**. He has identified three inputs, three processes and two outputs.

In Figure **C** another student, Jo, has carried out a more detailed analysis for her project, which is an electronic pet. She has started by

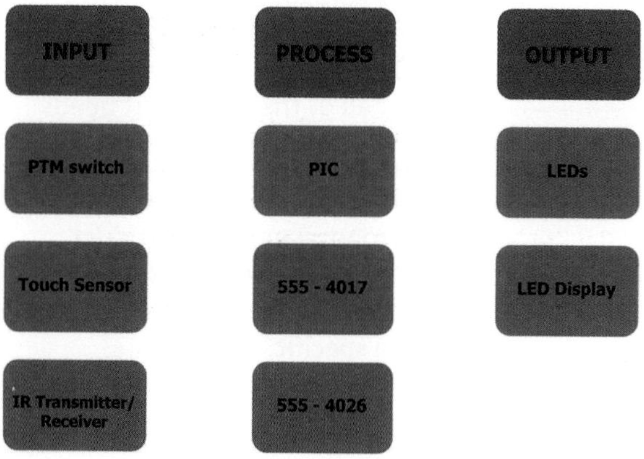

B *Systems analysis for a timing system by Jaymil*

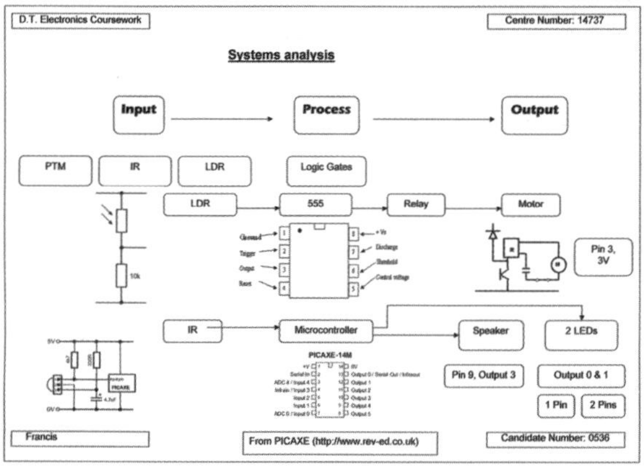

C *Systems analysis for an electronic pet by Jo*

analysing the process blocks that she needs to consider to come up with an electronics solution. She has then started to develop a solution by identifying particular components that could be used.

Tom has produced an outstanding systems analysis for his project, which is focused on a counting task for a specific application, Figure **D**. He has generated a vast number of possible systems from which to choose a solution. He has extended this to identify the electronic building blocks that he is going to use to develop the circuit, Figure **E**.

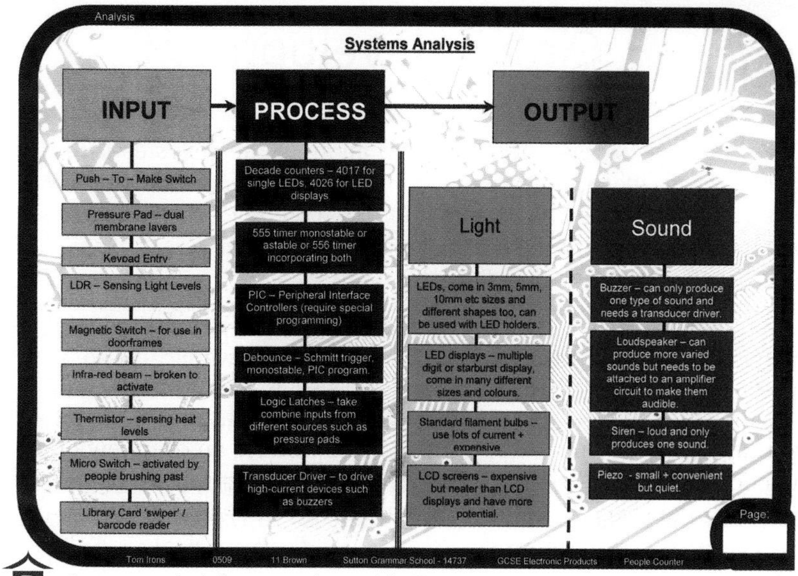

D *Systems analysis for a counting task by Tom*

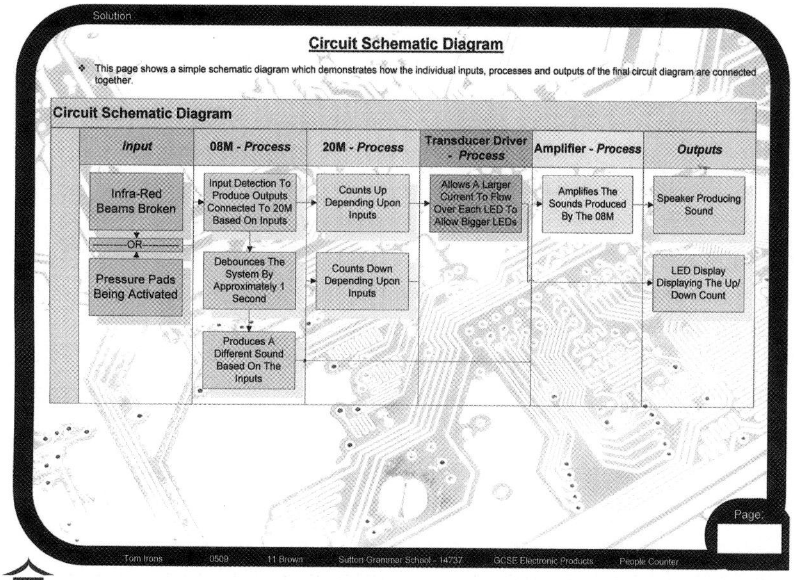

E *Identifying the electronic building blocks for a counting task by Tom*

Activities

1 Produce a systems diagram for a security light that is positioned outside a garage. It should be activated when an intruder comes near and should stay on for 15 minutes. Clearly label the signal, input, process and output.

2 Produce a systems diagram for a wireless doorbell which plays a selection of tunes when activated.

Summary

A system is a collection of parts that exist to form a function.

The parts of a system are signals, inputs, processes and outputs. A complex system may have several process boxes of each type.

Systems analysis is a useful tool to help identify the functional parts that need to be included in a product.

AQA Examiner's tip

Use systems analysis to generate a range of ideas from which you can identify a solution.

Use a systems diagram to work out the parts that need to be included in your design.

kerboodle!

8.5 Circuit development

Once you have carried out the systems analysis, the next step is to develop these ideas into actual circuits. You should present your circuit development by using a range of traditional methods such as sketching of circuits, or **virtual modelling** using computer software, or real-world modelling using breadboards.

Most people associate computer-aided design (CAD) with producing high-quality working drawings. However, CAD software can also be used during the development of electronic products to select suitable components for the working parts of the product and then model the function of the finished product. This saves the cost and time of having to get hold of all the components you would need to make a **real-world** model of the circuit. However, there is always the slight element of uncertainty about whether a virtual model will work when the real-world circuit is made.

Breadboarding on the other hand, although fiddly and sometimes unreliable, has the advantage of being a real-world model. If something works on a breadboard it has every chance of working as a completed piece of practical electronics.

It is probably best practice to sketch out on paper what you think your initial circuit may look like and then use a CAD program to model it. This shows your initial development and can save some time, by allowing any modifications needed to make a working circuit to be identified.

Figure **A** shows one example of Jaymil's development of his systems analysis into a possible electronic idea. An alternative way of presenting this, which links with more clearly the systems analysis, is shown in Figure **B**. However, only one of these options should be used. Both meet the assessment criteria.

Once you have generated a range of ideas, you should model your favourite option. Figure **C** shows a virtual model for a 'random lottery number generator', generated using CAD software. The student, Cho, has used a 555 astable to send pulses to the cascaded IC 4017 decade counters. The circuit has the facility to control the speed of the pulse and has a push-to-make switch to hold the decade counters,

Objectives

Explain two forms of circuit modelling.

Develop and model electronic ideas.

Key terms

Virtual modelling: a computer simulation of a system that enables a user to perform operations on the simulated system.

Real-world: actual experience or practice, as against a virtual or theoretical object.

AQA Examiner's tip

Remember: at this point you are still selecting and rejecting ideas. Do not throw any sketches away or delete any work from the computer.

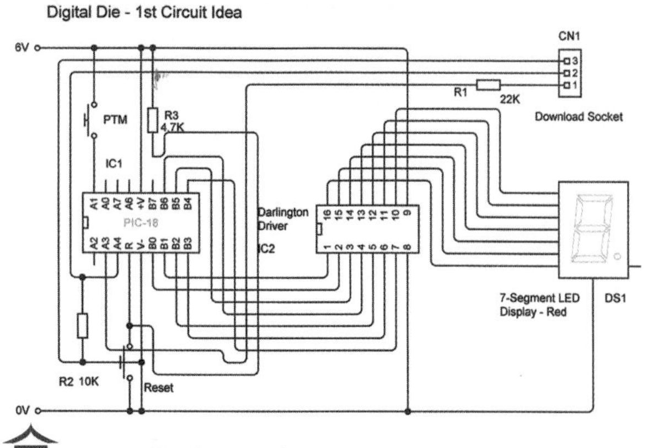

A *Circuit ideas by Jaymil*

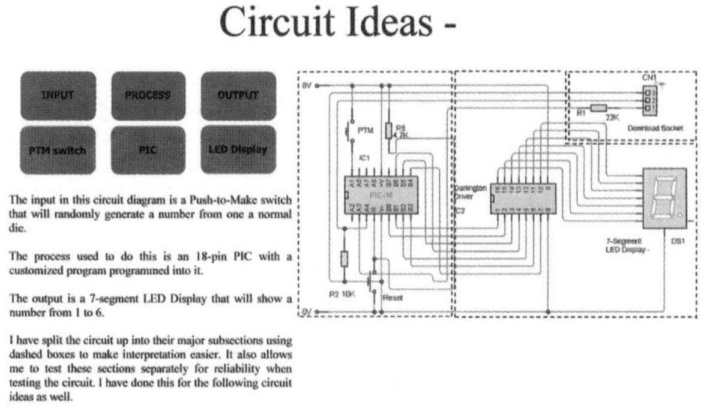

B *Circuit idea showing systems blocks*

to generate a random number. The success of the circuit depends upon a very high pulse speed.

Before you commit to designing a PCB, you should make sure that the circuit works in the real world, by making a model using a breadboard with real components. Jon was developing a remote control buggy for his project and built the breadboard shown in Figure **D**. He is modelling using an 18-pin PIC, sending a signal to an L293D driver to control the buggy motors. The 18-pin PIC receives IR signals from an 8-pin PIC handset.

A final piece of circuit development, and a good example of quality control, is shown in Figure **E**. To check that all the components that worked successfully on the breadboard are included in, and fit onto, the PCB design, Tom has put a printout of the PCB design on a piece of foam, in this case an upturned mouse mat. The components are stuck into the foam through the paper. When you have satisfactorily completed the check, keep the sheet used to provide evidence of quality checking for your folder.

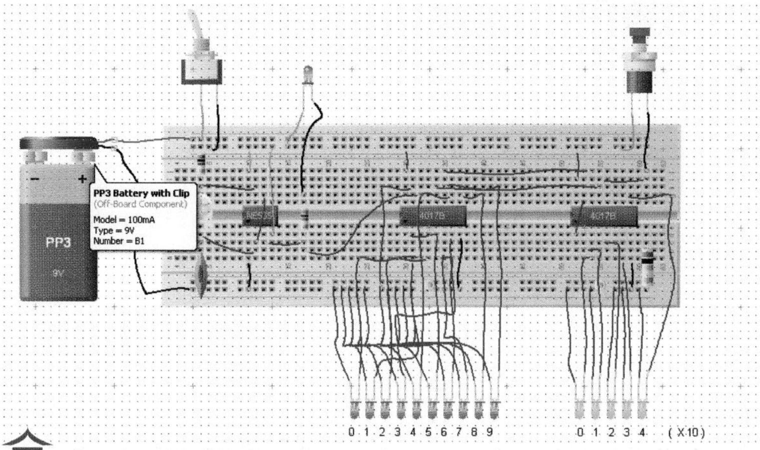

C *Virtual model of Cho's random number generator*

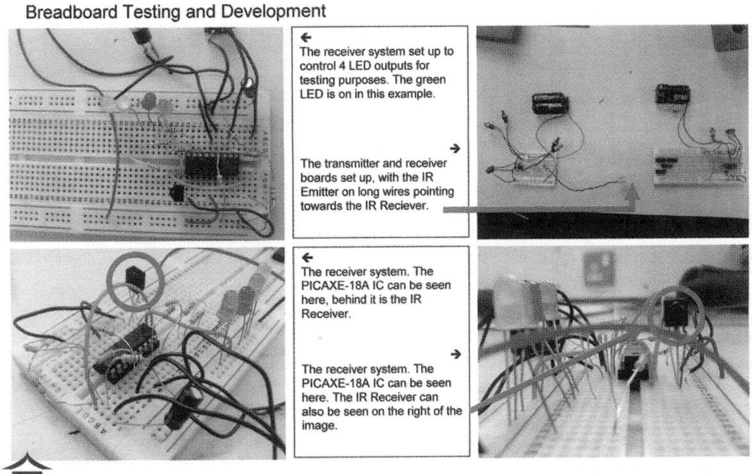

Breadboard Testing and Development
The receiver system set up to control 4 LED outputs for testing purposes. The green LED is on in this example.
The transmitter and receiver boards set up, with the IR Emitter on long wires pointing towards the IR Reciever.
The receiver system. The PICAXE-18A IC can be seen here, behind it is the IR Receiver.
The receiver system. The PICAXE-18A IC can be seen here. The IR Receiver can also be seen on the right of the image.

D *Breadboard model of a circuit by Jon*

E *Assembling components on the PCB layout drawing by Tom*

Activity

The random number generator shown in Figure **C** could also be created using a PIC microcontroller as the input to the decade counters. Using the CAD software available in your school, produce a virtual model of this circuit. Produce a breadboard model of this circuit and compare the two models.

Summary

The initial ideas for the circuit can be developed by modelling. This can be carried out by sketching the ideas, CAD modelling and simulated testing, and real-world modelling using a breadboard.

AQA Examiner's commentary

Jaymil has used his systems analysis and produced three circuit diagrams from which he will choose his final solution Figure **A**. He has used a CAD program to draw these circuits, which will allow him to simulate the circuit and check whether it works. He has a good range of inputs, processes and outputs, but he will need to justify his final choice.

8.6 Designing the printed circuit board

Once the circuit design has been selected, the next step is to develop the PCB that the circuit will be assembled on. To explain how this is carried out, we will first look at how a PCB design is created manually on a computer, then how CAD software can be used to create a PCB design.

Throughout the process, you must bear in mind that when designing a PCB, it must meet any size requirements stated in the specification and fit inside the enclosure.

Objectives

- Explain the advantages and disadvantages of using computer software to develop PCB designs.
- Design a PCB for a circuit.

Creating a PCB design manually

We are going to create a PCB layout for the timer circuit to run a siren used in a keypad operated alarm. This is based around a monostable timer with a 555 IC, Figure **A**.

The schematic of the circuit may bear little resemblance to the assembled real-world circuit. One reason for this is that the IC pins in the circuit diagram are not in order. They are drawn that way to make it easier to break the circuit down into the input, process and output. Figure **B** shows a real-world IC. The pins are numbered from 1 to 8, anti-clockwise from the pin to the left of the notch or recess. There is usually a dot next to pin 1.

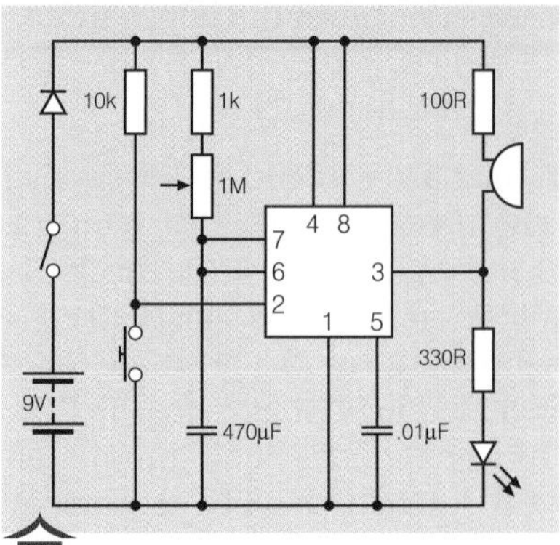

A *Schematic drawing of a monostable timer circuit*

The steps to produce a PCB **layout** for this circuit are:

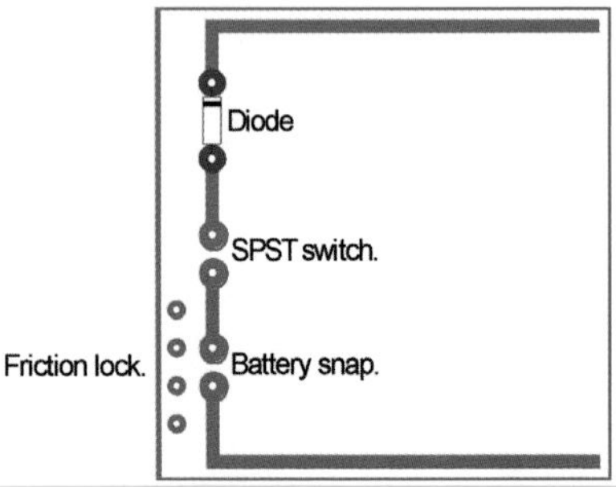

1 *Add pads for the battery, on-off switch and a diode. Connect all of the components with thick tracks and extend them to form a supply rail and a 0 V rail.*

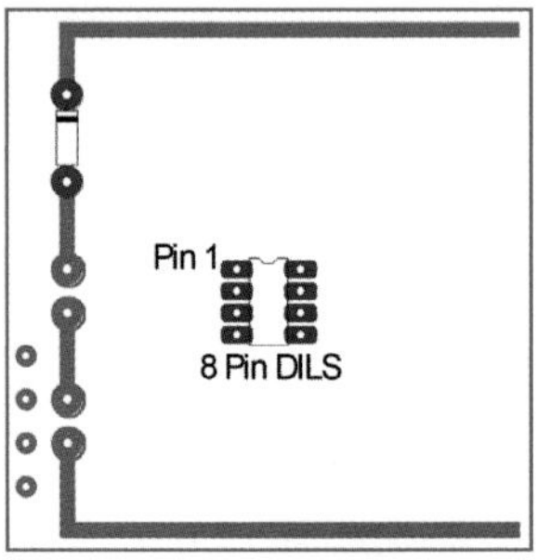

2 *Add the IC, ensuring that the notch between pins 1 and 8 is pointing upwards.*

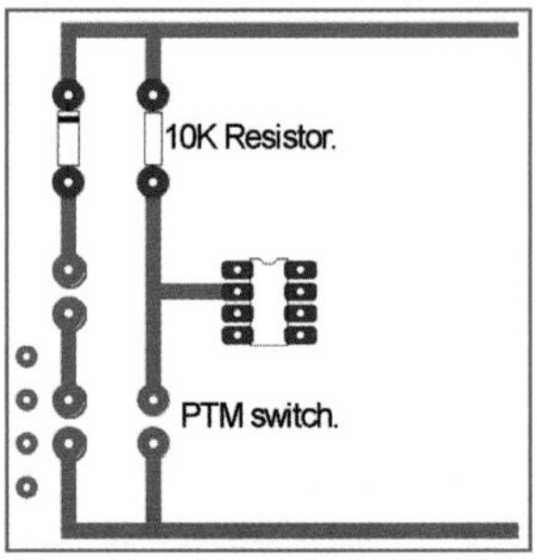

3 *Connect pin 2 up to the supply rail with a 10 k resistor. Connect pin 2 down to the 0 V rail with two pads that will carry the push-to-make (reset) switch.*

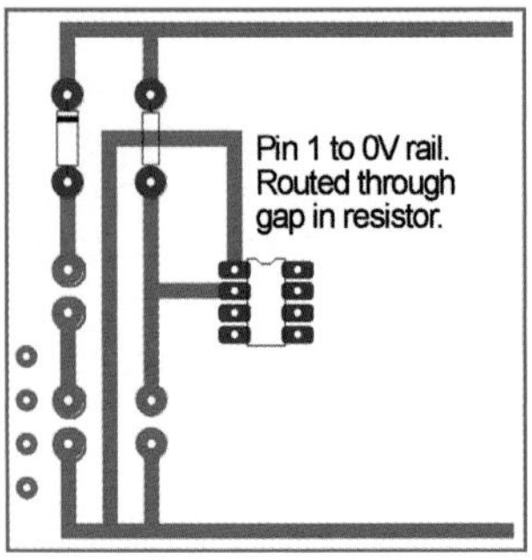

4 *Pin 1 can now be routed through the gap between the pads of the 10 k resistor..*

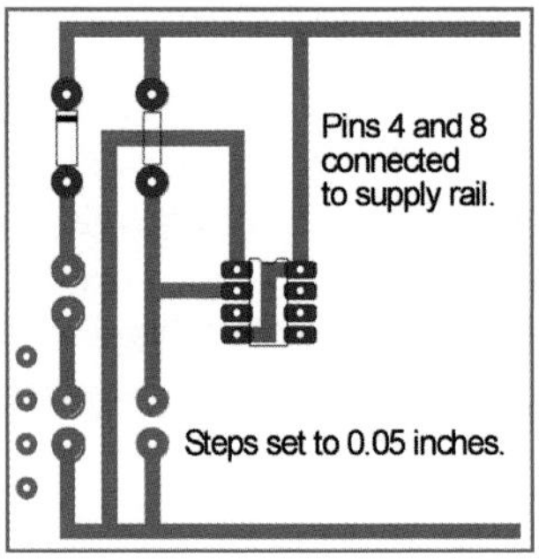

5 *Connect pin 4 to pin 8 and then to the supply rail.*

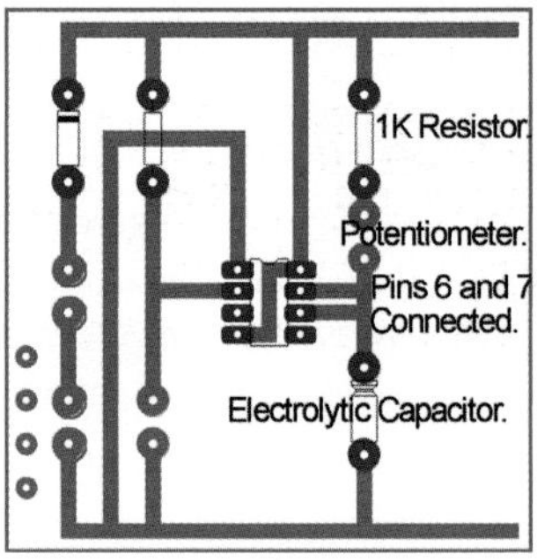

6 *Connect pins 6 and 7 to the 0 V rail through an electrolytic capacitor. They connect up to the supply rail through a potentiometer and a 1 k resistor.*

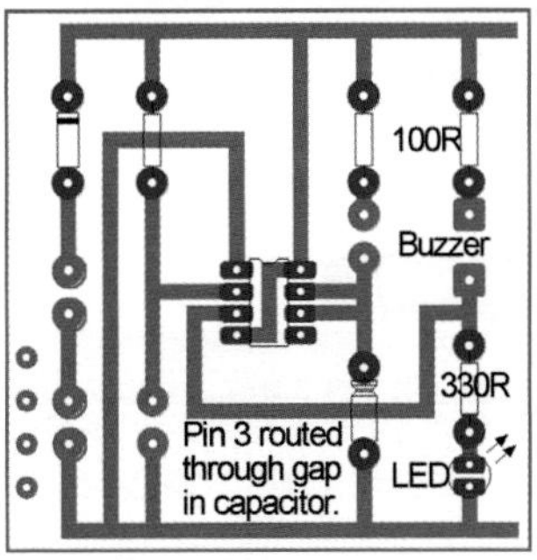

7 Connect pin 3 to the supply rail through the buzzer and 100 R resistor and down to the 0 V rail through the 330 R resistor and the LED.

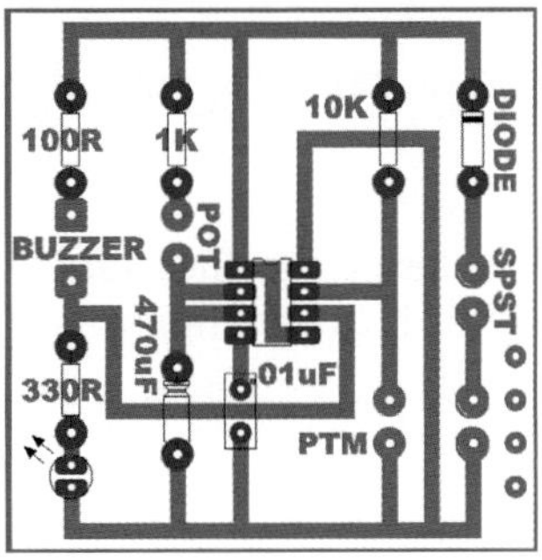

8 Connect pin 5 to the 0 V rail through a 0.01 µF capacitor. Delete any unwanted track. Place any text wanted on the board, such as component identifications. Add a border around the PCB, to provide a nice sharp line to trim the PCB back to after it has been produced. As a final check, some PCB CAD software will allow a PCB layout to be tested on screen.

Creating a PCB using CAD software

Many CAD software packages that design electronic circuits are able to produce a PCB layout at the push of a button. This can save a lot of time, especially for complicated PCBs that contain several ICs. However, CAD generated designs are often comparatively large and they may contain a high number of flying leads, where extra wires are needed to link tracks. Figure **D** shows PCB layouts for a monostable timer circuit produced (a) by a CAD package and (b) by a student. The student's layout, which works equally well, is much smaller than the layout generated using the CAD software. CAD generated design also often result in thin tracks close together, making it difficult to solder.

One effective way of using CAD software to create a PCB layout is to generate an initial layout using the software and then to customise and improve this layout manually. A good way of presenting this is to include a series of layout diagrams showing progressive changes, with an explanation of the changes made. This allows you to show both the use of CAD in your design folder and the application of your skills to the development.

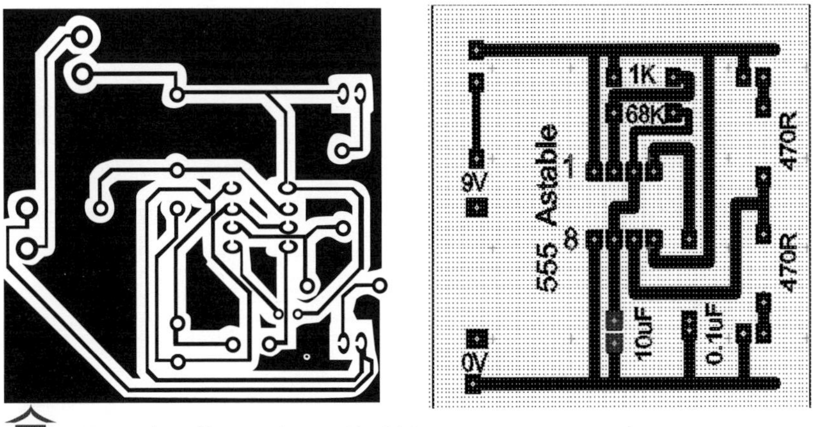

B *PCBs produced by a student and by CAD*

Activities

Using the CAD software available in your school, design a PCB for an astable 555 timer circuit.

Key terms

Layout: the design of the tracks on a PCB.

AQA Examiner's tip

You should include the different steps of developing your PCB in your design folder. Even if you create the original design using CAD software, you should customise it and improve it for your product.

Summary

CAD software can produce PCB designs from systems diagrams or circuit diagrams. However, this may not necessarily be the optimum size, shape and layout for the PCB.

Show the development of the PCB design in your folder.

8.7 Designing the enclosure – sketching

As well as designing the circuit, we need to design the enclosure or case that houses it. The first step in doing this is to generate some ideas for the case and capture them using sketching.

Objectives

- Explain how sketching is used.
- Render a sketch, using different line thicknesses and shading.

What is sketching?

The aim of **sketching** is to **communicate ideas**. Sketches often have lots of labels to explain or point out interesting features.

Sketches are often used to get initial ideas down on paper, for example as shown in Figure **A**. This is sometimes known as capturing ideas or producing concept drawings. Freehand sketching doesn't need any drawing equipment apart from a pencil or pen. The sketch should be produced quite quickly. However, this doesn't mean that it is rushed or unclear.

The standard of sketching varies a lot between different designers and different companies. Unlike most other forms of drawing used in technology, there are no conventions for producing sketches. They do not have to be produced to **scale**. Sketches can be 3D or 2D. 3D sketches are often used to show the whole object. 2D sketches are often used to show close-up views of individual features or details.

Key terms

Sketching: a quickly produced visual representation of an idea.

Communicate ideas: share a concept with others.

Scale: the ratio of the size of the drawing to the size of the part.

Rendering: applying colour or texture to a sketch or drawing.

Shading: creating different tones on a sketch or drawing.

Rendering

Rendering means adding colour or texture to a picture. Although sketches are produced quickly, they are often rendered to give a more realistic view of what the product could look like. Two common forms of rendering are thick and thin lines and **shading**.

Thick and thin lines

Different line thicknesses can be used to make parts of illustrations stand out. A simple, easy to use technique is:

1. The sketch is produced as normal, using thin lines to show where any two surfaces meet, Figure **B(a)**.
2. On any edge where only one surface is seen, the line is then increased to medium thickness, Figure **B(b)**.
3. On any outside edge, the line is made even thicker, Figure **B(c)**.

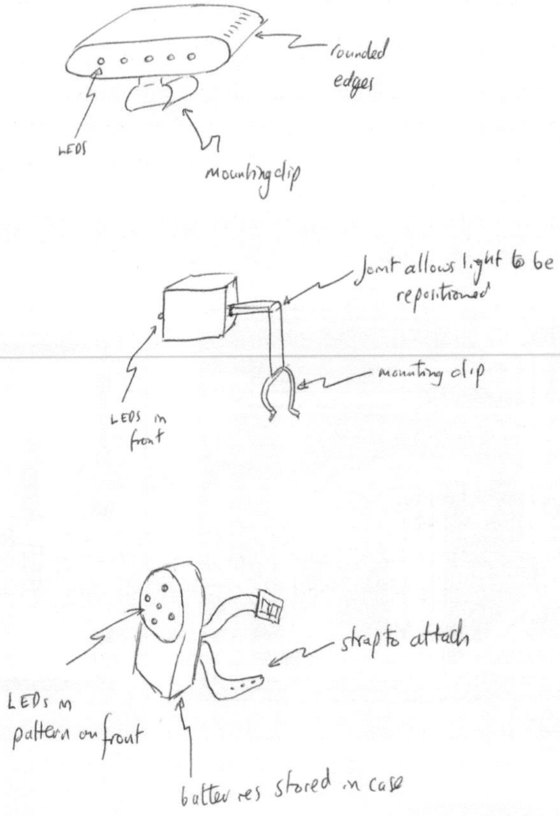

A *Examples of sketches: ideas generation for an enclosure*

This is easy to do if sketching is carried out in pen, as technical pens are available in different sizes, such as 0.25 mm, 0.5 mm and 1 mm. It can also be done in pencil, but care is needed to keep the line thickness consistent.

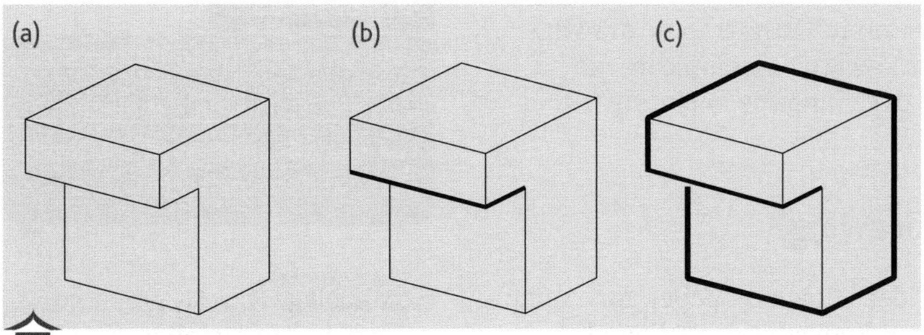

B *Thick and thin lines*

Shading

If an object is placed near a window or light, the side facing the window will appear to be a lighter colour than the side which is in shade. These lighter and darker versions of the same colour are called tones.

An object with one tone all over looks flat and uninteresting. A sketch shaded with different tones looks more realistic. With round objects, like balls or pipes, it also helps to make the surface appear curved.

Shading can be easily carried out using a pencil. For example:

1 Lightly shade the whole object, Figure **C (a)**.
2 For any sides or areas that are further from the light source or only receive light indirectly, shade them again, so that they become a bit darker, Figure **C (b)**.
3 For any sides that have no direct exposure to the light source, shade them again, so that they become even darker.

Shading is normally most effective if a single colour is used, with different tones. Using too many different colours can reduce its impact.

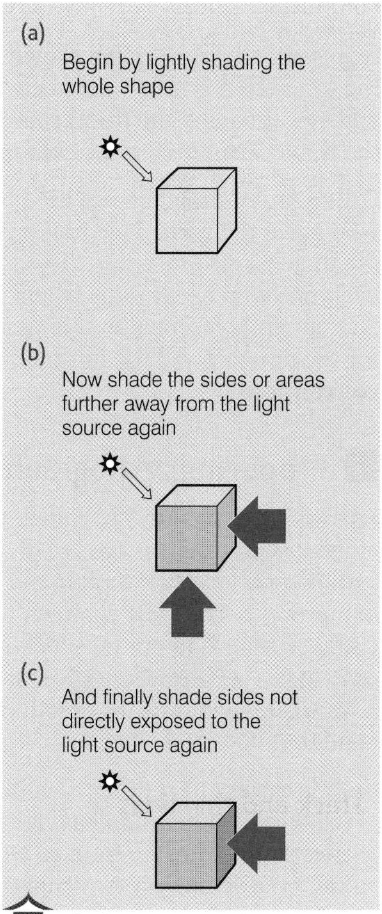

(a)
Begin by lightly shading the whole shape
(b)
Now shade the sides or areas further away from the light source again
(c)
And finally shade sides not directly exposed to the light source again

C *Shading*

Activity

Produce a series of quick sketches of a television. On different versions of the sketch, use thick and thin lines and shading to see how these make it look more realistic.

AQA Examiner's tip

Master a few simple rendering techniques, such as controlling line width and shading, to make your sketches look more realistic.

Summary

Sketching is a method of communicating ideas. There are no conventions for producing a sketch.

Sketches and drawings can be made to look more realistic by rendering. Rendering techniques include controlling line width and shading.

kerboodle!

8.8 Designing the enclosure: CAD drawings

Sketches normally do not contain enough information to make a part, such as the dimensions of every feature on it. Orthographic drawing is a formal drawing technique which is used to communicate the dimensions of a part. The next step in designing the enclosure is to produce an orthographic drawing of it.

Objectives

Explain what the function of an orthographic drawing is.

List the advantages of CAD over manual drawing techniques.

What is an orthographic drawing?

An **orthographic drawing**, also called a working drawing, usually shows three 2D views of a part. There are several different possible combinations of views. However, the most common form of orthographic drawing is called third angle projection. This is represented by the symbol shown in Diagram **A**. It shows the part from the top, which is called the plan elevation, the side, which is called the end elevation and the front, called the front elevation. On the drawing, the three elevations are laid out in an L-shape, as shown in Diagram **B**.

Orthographic drawings are used by designers, engineers and the people who make the parts. The drawings are produced to scale and show the dimensions of the part. To ensure that they can be understood by everyone who needs to use them, it is important that they are always laid out and presented in the same way. These 'rules of presentation' for the drawing and the information it contains are called drawing **conventions**.

Key terms

Orthographic drawing: a working drawing of a part showing three views, to communicate the dimensions of the design.

Conventions: rules of presentation that drawings must conform to.

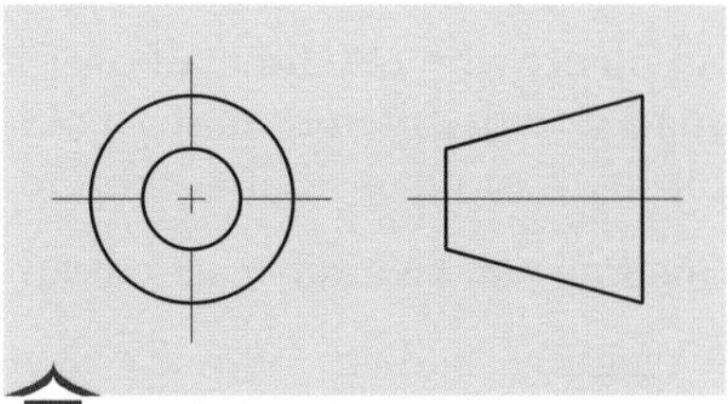

A *Symbol for third angle projection*

Producing orthographic drawings using CAD

As explained in topic 8.6, computer-aided design (CAD) is the use of software packages to assist in the design of a product. The most common use of CAD software is to produce high-quality drawings. CAD drawings have many advantages over drawings produced by hand:

- It is easier and quicker to make changes. Rather than having to restart the drawing from scratch, a CAD drawing can be changed by editing the existing file.
- They can be more accurate.
- They can be saved electronically, saving space. They can be easily circulated to anyone who needs them, using CDs or e-mail.
- Some parts, such as screws, nuts and bolts, can be downloaded from libraries of CAD parts. This means that they don't have to be drawn, saving lots of time.

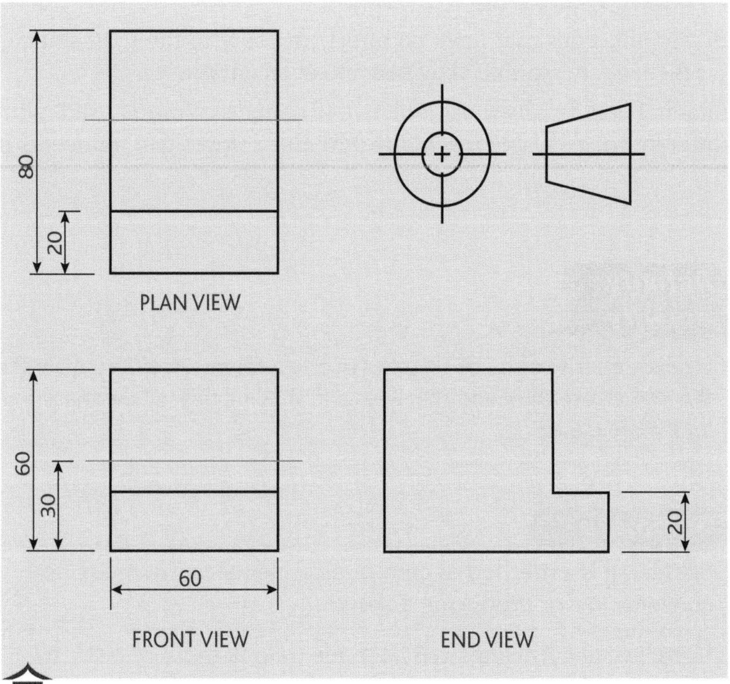

B *Third angle projection*

There are two main types of CAD drawing software: two-dimensional (2D) and three-dimensional (3D). Both types of software normally include a wide range of drawing tools. These are the basic commands used to create CAD drawings. The tools typically range from drawing simple lines and inserting shapes, through to duplicating and manipulating the drawn features and modifying and deleting features.

2D CAD

Drawing using 2D CAD software is similar to drawing by hand, Figure **C**. The screen has a working area, which is in effect the piece of paper that you draw on. This area can be changed to almost any size, and you can zoom in to see features close up.

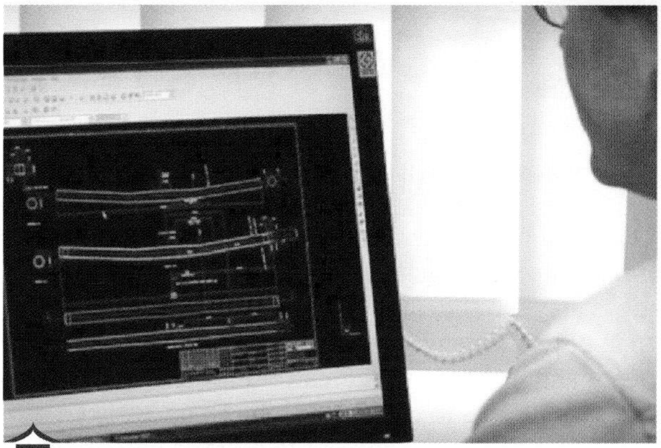

C *Creating a 2D CAD drawing*

3D CAD

In 3D CAD, what is created is a three-dimensional model of the part being drawn, Figure **D**. Most 3D CAD software also has the ability to produce working drawings directly from the CAD model.

Starting to draw a 3D shape is more complicated than drawing a 2D shape. In effect, you have an empty three-dimensional space that you will fill with the design. You first have to create an imaginary piece of 2D paper within this space. This is often called a plane or workplane. The design starts as a 2D drawing on this workplane. Once this has been created, it can be extruded to give it thickness or depth. Other features can then be added to this design. In addition to being able to change the work area and zoom in or out, the design can also be rotated and moved on the screen, so that it can be viewed and edited from any direction.

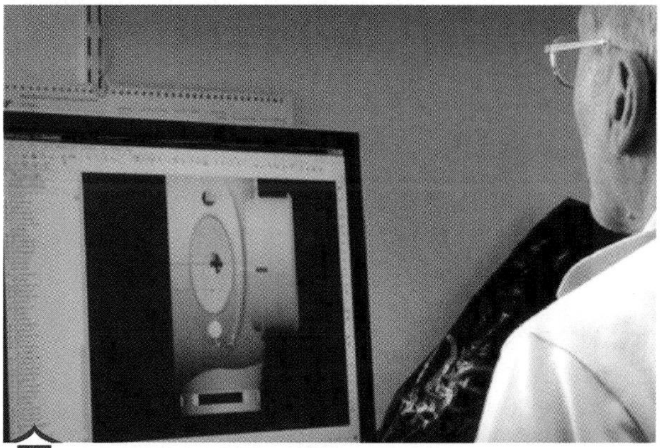

D *Creating a 3D CAD model*

AQA Examiner's tip

Use CAD drawings to demonstrate your ideas and, if possible, to output the information to a computerised machine.

Remember

Blueprints

Working drawings are sometimes also called blueprints. This term dates back to when all drawings were done by hand. As the drawing was produced, there was a sheet of blue carbon paper underneath it that the design also appeared on. This blueprint of the design was used on special equipment to print copies of the drawing.

Activity

Using the CAD software available in your school, produce a drawing of a lunar buggy that could be used to drive two astronauts around on the surface of the moon.

Summary

The purpose of an orthographic drawing, also called a working drawing, is to communicate the dimensions of a part.

Drawings produced using CAD software can be quicker to change and more accurate than drawings produced manually.

kerboodle!

8.9 Evaluating your design ideas

Choosing which design ideas to develop

Once you have generated some ideas for your circuit and enclosure, you have to decide which one to develop into your design proposal. Some ideas may have greater potential to meet the needs of the target market than others. One way of evaluating your ideas is to compare them to the needs shown in the **specification**, for example Table **A**. At this stage, you may not be able to evaluate all the characteristics of the design. This can mean that the evaluation of the design's potential may need to be based on subjective opinions about existing, similar designs.

Objectives

Explain why it is important to compare design ideas to the design needs.

Compare your ideas to the specification.

A *Example of an evaluation against the needs listed in the specification for ideas for a keypad operated alarm*

Spec. point number	How evaluated	Design A	Design B	Design C
1	Checking the CAD model of the circuit	Yes	Yes	Yes
2	Looking at existing products and checking this using the CAD models of the circuit	Yes	Yes	No
3	Checking the CAD model of the circuit	Yes	Yes	Yes
4	Checking the CAD model of the circuit	Yes	Yes	Yes
5	Checking the CAD model of the circuit	Yes	Yes	Yes
6	Testing existing products	Yes	Yes	Yes
7	Examining existing products	Yes	Yes	Yes
8	Looking at existing products and checking this using the CAD models of the circuit	Yes	Yes	No
9	Examining existing products and checking the CAD model of the design	Probably	Probably	Probably
10	Examining existing products and checking the CAD model of the design	Yes	Yes	Yes
11	Looking at similar products	Yes	Yes	Yes
12	Looking at similar products	Probably	Probably	Probably
13	Carrying out a silk test on similar products	Probably	Probably	Probably
14	Examining similar products and getting opinions from my teacher and the technician	Yes	Yes	Yes
15	Adding up the cost of the parts used in the CAD model	Probably	Yes	Probably

Note: Specifications needs would normally be listed, but have been omitted for presentation purposes

Key terms

Specification: a list of needs that the product must satisfy.

Ideally, you should choose to develop a design that does not have any 'No's' and has the potential to meet all the needs of the specification. If designs have similar ratings, you may need to make a choice. This could be based on how easy different designs would be to make, their comparative cost or any additional features that they may offer. It is important that you explain this decision in your design folder.

AQA Examiner's tip

In your folder, you should also use labels and comments on your design ideas to show that you are considering how well the ideas meet the needs of the target market.

Evaluating your design proposal

Once you have developed a design proposal for the final product, before you make it you need to be as certain as you can be that it will meet all the needs of the target market. This can be carried out by making an objective comparison against the specification, for example Table **B**.

Most of this evaluation can be based on testing any models made during the design process. A number of different tests will probably need to be carried out. The testing should be objective; this means that it should be based on facts and numbers, rather than opinions. Where this is not possible, sometimes it is necessary to state that some features will need to be further evaluated after a prototype is made.

Activity

Create a table listing all of the different needs that you might need to test when designing an MP3 player. For each need, identify how it could be objectively tested.

B *Example of an evaluation against the specification for the final proposal for a keypad operated alarm*

Spec. point number	Need	How evaluated	Pass?
1	The product is a keypad controlled alarm system. It requires a 3-digit code number to be entered to turn off the alarm.	Checking the CAD models of the circuit, and that the breadboard model works	Yes
2	The product should be controlled by a picaxe-20M microcontroller.	Checking that the breadboard model works	Yes
3	The product should be linked to two contact sensors and two motion sensors, using multi-core wire.	Checking that the breadboard model works	Yes
4	It should include five LEDs, to show whether the alarm is turned on and which sensor has been activated.	Checking that the breadboard model works	Yes
5	The control box for the product should be powered by three 1.5 V AA batteries and include the keypad.	Checking that the breadboard model works	Yes
6	It should include a relay to operate a siren, which can be heard up to 30 metres away.	Testing that the siren worked while standing 30 metres away	Yes
7	The siren should be housed in a separate enclosure and powered by four D batteries.	Testing that the siren worked with a pulsed signal	Yes
8	It should include an on-off switch and a download socket, so that it can be reprogrammed.	Checking that the breadboard model works	Yes
9	The enclosure for the control box should be made of white high-impact polystyrene, with a maximum size of 150 × 100 × 25 mm.	Measuring with an engineering rule; dimensions 140 × 95 × 25 mm	Yes
10	The enclosure for the siren should be made of yellow HIPS, with a maximum size of 150 × 250 × 100 mm.	Measuring the model with an engineering rule; dimensions 140 × 200 × 94 mm	Yes
11	The PCB, siren and batteries must be secured within the enclosures so that they cannot move around.	Looking at similar products	Yes
12	It should be possible to open both enclosures to change the batteries. However, it must be possible to close them to prevent water entering.	Testing the model of the case	Yes
13	There must be no sharp edges on the enclosures.	Carrying out a silk test on the model of the case	Yes
14	The product must be suitable for one-off production.	Making models and prototyping the circuit on a breadboard	Yes
15	The total cost of the parts used must be less than £10.00.	Adding up the cost of the parts used on the materials list; total £8.12	Yes

Summary

You should evaluate your design ideas against the specification and design proposal.

kerboodle!

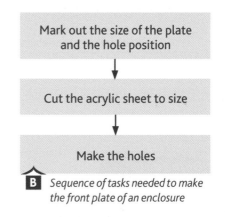

8.10 Preparing for production – the production plan

The role of the production plan

A **production plan** is a set of instructions for making the product. It should contain enough information that someone who has never seen the finished product should be able to make it. The main reasons for preparing a production plan are shown in Table **A**.

A *Reasons for preparing a production plan*

Type of benefit	Reason
Quality control	It helps to prevent design features being missed during manufacture
Consistency	If the same product has to be made again in the future, it can be made the same way
Safe working	The production planner will identify any risks that might arise during making and will take action to reduce them. It also helps the people making the part to be aware of how to work safely
Efficient use of machines	It helps the company to plan the use of machines. This can be very important in a busy workshop when many different parts are made on the same machine
Financial management	The required amount of materials can be planned, so that they can be bought for when they are needed, rather than having to take up expensive storage space and have funds tied up in unnecessary items

The production plan for your project will need to cover the circuit (including the PCB) and the enclosure. For many projects, making the enclosure may need a wider range of different process types than making the circuit, so this will be explained first.

Planning to make the enclosure

The first task is to look at the design of the enclosure, see what materials it is to be made from and then to identify which processes need to be carried out to turn the materials into the finished part. For example, consider a simple rectangular front plate for a case, made from acrylic sheet, with a hole for a screw at each corner. If a large sheet of acrylic was available to make it, you might identify that you need to make it out, make the holes and cut it to size.

Next, the tasks need to be put in a suitable order. Some tasks may need to be carried out before others are possible. For example, for the front plate, marking out needs to be carried out before cutting. This sequencing can be shown as a flowchart, Figure **B**.

On its own, the sequence of tasks does not give enough information to make the product. The other things that the person making the product needs to know now need to be shown in the production plan, Table **C**.

Objectives

Explain why it is important to prepare a production plan.

Be able to prepare a production plan.

Remember

Information needed in the production plan

The tasks to be carried out

Time needed for each task

Materials and parts to use

Tools and equipment to use

Quality checks to carry out

Safety notes

Links to the specification

Tools to use if the first choice is not available

Key terms

Production plan: the instructions on how to manufacture a product.

AQA Examiner's tip

Break the processes into small operations and make sure that the sequences make sense.

Ensure your plans have sufficient detail for a third party to be able to use them.

Mark out the size of the plate and the hole position

↓

Cut the acrylic sheet to size

↓

Make the holes

B *Sequence of tasks needed to make the front plate of an enclosure*

C *Extract from a production plan for the manufacture of the front plate of an enclosure*

Step	Task	Links to spec	Time, minutes	Tools to use:	Materials to use:	Quality control:	Safety notes:
1	Mark out the size of the plate and the centre points of the holes See working drawing for dimensions	1, 2, 3, 11, 13, 14	45	Scribe, engineering rule	Acrylic sheet, 1000 × 1000 × 2 mm	Check the sizes using an engineering rule Check the hole positions using the template. If wrong, repeat task	Wear gloves when handling the sheet, as it may have sharp edges
2	Cut the plate to the size needed	12, 14	5	Scroll saw	Acrylic sheet 1000 × 1000 × 2 mm	Check the dimensions using an engineering rule	Use the machine guard. Wear goggles. Wear gloves when handling the large sheet.
3	Make the holes in the plate See working drawing for the drill bit sizes to use	1, 2, 14	20	Pillar drill, HSS drill bits	Front plate from task 2	Check the hole diameters using Vernier callipers. Check the hole positions using the template. If holes too small, re-drill. If holes too big or in the wrong position, go back to step 1	Use the machine guard and wear goggles. Use a machine vice to hold the plate securely

Planning to make the circuit

The planning for making the PCB can be carried out using the same format as above. As making the circuit uses a single process it is sometimes included only as a single line on this format of plan. However, because of their size, shape and position, it may be necessary to attach and solder components in a certain sequence. This can be covered by producing a separate list giving the sequence for attaching components, for example Figure **D**.

Activity

Create a production plan for the manufacture of a chair. The seat and back should be made from thermoplastic, available as granules. The legs should be made from steel tubes. The legs will be attached to the seat and back using screws, which are available as standard parts.

Instructions for Making the Alarm Circuit

1. Place the 1 k (brown black red gold) and 22 k (red red orange gold) resistors in position. Bend the legs to hold the resistors in position and then solder.
2. Place the two 10 k (brown black orange gold) and 330 (orange orange brown gold) resistors in position. Bend the legs to hold the resistors in position and then solder.
3. Solder the 1N4001 diode is position D1, making sure the silver bar is pointing towards the right hand side of the PCB.
4. Push the download socket onto the PCB and make sure it clicks into position (so that it lies flat on the board). Solder the five metal square contacts (the five round plastic support post holes do not have to be soldered).
5. Push the IC socket into position. Make sure the notch at one end points up towards the socket.
6. Solder the BC548B transistor in position, making sure the flat side aligns with the marking on the PCB.
7. Solder the 100µF capacitor is position, making sure the + and – legs are around the correct way.
8. Thread the battery clip down through the large hole by the letters DER. Thread it back up through the large hole by the letters ORD. Then solder the black wire in the hole marked 0V and the red wire in the hole marked V+.
9. Carefully check the board to make sure there are no missed joints or accidental solder bridges.

D *Instructions for soldering an alarm circuit*

Summary

The production plan provides detailed instructions on how to make the product. It should list the tasks to be carried out, in the correct order along with all the other information needed to make the part.

kerboodle!

8.11 Risk assessment

What is a risk assessment?

A **risk assessment** identifies any potential **hazards** associated with an activity. This means that it lists any ways that the activity could cause harm. It also lists the risk mitigation actions. These are the things that need to be done to make sure that any risks identified are reduced as much as possible.

Risk assessments are carried out for a wide range of activities, ranging from manufacturing to sports to travel. Here we will focus only on their use during the making of electrical circuits and the manufacture of the cases or enclosures for electronic products. You should carry out a risk analysis for every process listed on your production plan.

The aim of the risk assessment is to reduce the risk of harm to:

- the people carrying out the activity
- other people who may be affected, such as other students, teachers and visitors
- the working environment.

Risk assessments should be prepared for every activity carried out during the manufacture of a product. These should be practical documents that are used by the people carrying out the operations to help make sure that they work safely.

Objectives

- Understand why risk assessment is important.
- Carry out a risk assessment.
- List a range of mitigation actions that could be stated on a risk assessment.

Key terms

Risk assessment: a review of the potential of an activity to cause harm.

Hazards: things that cause a risk of harm or injury.

Mitigation actions: precautions taken to reduce a hazard.

How to carry out a risk assessment

The first step in carrying out a risk assessment is to identify possible sources of harm. For example, during a machining operation to make an enclosure for an electronic product, the following questions might be asked:

- Does the machine have rotating or moving parts that the operator could get caught in?
- Does the process produce any debris? How would this affect the user, nearby people and the environment?
- How noisy is the process?
- What happens if the tool breaks or shatters during the process?
- Are there any features of the work piece that could cause harm?
- Is the work piece fixed in place during the operation or could it become a hazard if it moves?
- Is the process or work piece at a temperature that could cause harm?

The next step is to rate each of these risks, by their potential impact and how likely they are to occur. One way of doing this is to use a high, medium and low rating system. For example, a high impact shows that it could cause a lot of harm, whereas a low impact shows that any resulting injury would be minor. It is important to remember that this type of rating is subjective and will vary between different uses of the same equipment. Table **A** shows an example of this for the manual soldering of a circuit.

A *Example of a risk assessment for soldering*

Risk	Who this affects	Impact	Likelihood	Mitigation actions
1. The tip of the soldering iron is very hot. This can cause burns to hands and scorch other objects it touches.	User, equipment used in the working environment	Medium	High	Always assume that the soldering iron is hot – never touch the hot end or you might get burnt. Never touch anything with the hot end except what you are soldering. Always use a baseboard and a stand for the soldering iron, so that the work surface does not get burnt.
2. Heat can also be conducted down the solder wire, causing burns.	User	Low	Medium	Never use a piece of solder less than 5 cm long. This stops your fingers getting burnt by heat passing down the wire.
3. Soldering makes smoke called soldering fume. This is poisonous and can make you ill.	User, nearby workers, the environment	High	Low	Use a fume extractor to take away the fume.

Risk mitigation actions

Once the risks have been identified, the next step is to identify how each of them can be reduced. There may be one or a number of actions for each risk. There is a broad range of possible **mitigation actions**, such as:

- using equipment such as machine guards or fume extractors
- following a written safe working procedure to use the process or having specialist training to use it
- using appropriate Personal Protective Equipment (PPE), such as goggles, gloves, ear defenders, breathing masks or aprons.

Table **A** includes some of the mitigation actions used when soldering manually. For comparison, Table **B** shows examples of mitigation actions that are commonly used during the drilling of sheet metal.

B *Examples of some hazards and mitigation actions during a drilling operation*

Hazard	Mitigation actions
Machine has moving parts	Use machine guard, tie back hair Secure any loose clothing, such as ties
Machining operation produces swarf	Wear safety goggles, wear an apron, wear thick gloves when removing swarf
Tool may break or shatter during the process	Use machine guard, wear safety goggles
Work piece has sharp edges	Wear gloves
Work piece is heavy	Use correct lifting technique, lift with someone else, use a crane to move it
Work piece might move during process	Use a machine vice or clamp to hold it in place
Work piece becomes hot during the process	Allow it to cool down before moving it

AQA Examiner's tip

It is essential that you can work safely with any process that you need to use.

Activities

1. Create a risk assessment for one of the following tools that you are familiar with: drill, CNC router or vacuum former.
2. Create a risk assessment for the manufacture of a PCB using chemical etching.

Summary

A risk assessment evaluates any potential hazards associated with the use of a tool or a piece of equipment.

The risk assessment should include mitigation actions to reduce the risks. These could include the use of fume extraction, machine guards, personal protective equipment and safe working procedures.

kerboodle!

8.12 Preparing for production – timing plans

Why is the timing plan important?

A timing plan has a different role to a production plan. A production plan is a complete, detailed set of instructions for making the part. A timing plan is used to plan when a sequence of activities will be carried out.

One of the most frequent uses of timing plans is for project management. They are used to manage the construction of new buildings, plan space missions and develop new products. The other main use of timing plans is to plan the manufacture of individual products or batches of products. In both cases, the most common format used is called a **Gantt chart**.

Timing plans allow realistic completion dates to be worked out for a sequence of activities. In industry or construction, these dates will normally be agreed with the customers. If a company fails to meet its agreed deadlines, customers may take future business elsewhere.

Creating a Gantt chart

Each step in the project or manufacturing process must be listed in the correct sequence, along with an estimate of how long it will take. Some of the steps may run concurrently – this means they could run at the same time or overlap other tasks. For example, Figure **A** shows the steps to make a cup of coffee. If you waited till each step finished before starting the next step, it would take a long time to make the coffee. If you put the coffee, sugar and milk in the cup whilst the water is being heated it is a lot quicker!

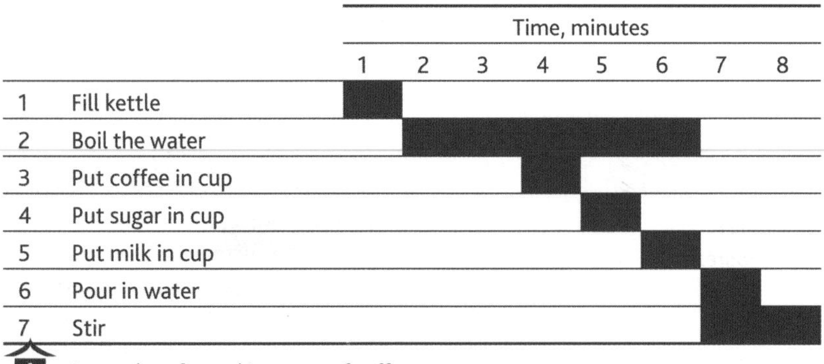

A *Gantt chart for making a cup of coffee*

Using the timing plan to monitor progress

There are many reasons why individual activities within a project or manufacturing sequence may take different times from the plan. For example, machines may break down, components may fail quality checks and need to be remanufactured, or material from suppliers may arrive earlier or later than planned. During your project, you can use your timing plan as a management tool to monitor whether the activities are on target to finish on time. You can mark on the plan to show when activities are being carried out and completed.

Objectives

- Understand why a timing plan is needed.
- Create a timing plan.

Key terms

Gantt chart: a type of timing plan.

Lead time: the amount of time needed to complete an activity or to supply a product.

Critical path: the shortest route through the timing plan, where each step contributes directly to the lead time.

Remember

Using a Gantt chart to work out completion dates

The shortest possible time in which all the activities can be completed is called the **lead time**. This is the time from the start of the first activity to the end of the last activity, taking into account any overlaps or concurrent activities.

The steps that make up this route through the plan are called the **critical path**. If any of these steps is delayed, the whole activity will finish later.

If activities are behind the schedule, actions can be carried out to speed them up. For example, in industry this may be the use of overtime to speed up manufacturing. If this happens in your project, it is a good idea to add labels on the plan, explaining what has caused you to fall behind schedule and what you did about it. This can provide good evidence of the work you carried out and the actions you took to address problems.

AQA Examiner's commentary

This is a good, very detailed timing plan for a project, which also shows the progress made. It could be improved by adding labels explaining why some of the steps were different to the planned timescale.

Example of a timing plan

Figure **B** shows a student's timing plan developed for a complete GCSE Project. This type of plan can be easily created using a table in a word-processor software package or a spreadsheet. A major challenge when creating a timing plan for the full project is that it may involve guesswork or assumption about the design and how it will be made. For this reason, in some projects the final timing plans the manufacture are not prepared until after the design has been chosen.

AQA *Examiner's tip*

Break the processes into small operations and make sure that the sequences make sense.

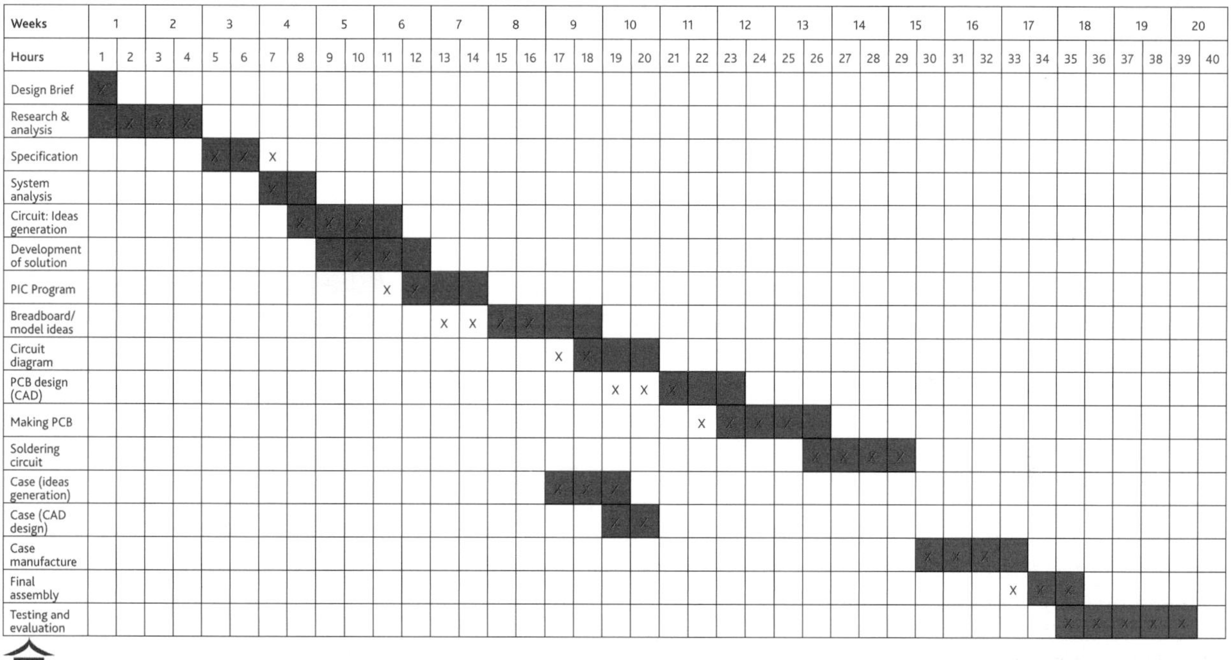

B *Gantt chart prepared by a student for a GCSE project*
Key: Red squares indicate plan. X's show actual progress.

Summary

A timing plan, such as a Gantt chart, is used to plan when a sequence of activities will be carried out and to monitor progress whilst the activities are being carried out.

It can be used to calculate realistic completion dates for an activity or the manufacture of a product.

Activity

Using Figure A for guidance on times, create a Gantt chart for making a cup of tea (with milk and sugar). If you had to make a cup of tea at the same time as the cup of coffee, identify any activities that might clash and re-plan them accordingly. What would be the lead time to make the two drinks separately? What would be the lead time to make the two drinks together?

kerboodle!

8.13 Making – high-level skills and safe working

Working safely

The practical work that you will carry out forms an important part of your assessment in GCSE Electronic Products. You will be assessed both on how well you work and on the standard of work you produce.

Make sure that you know how to carry out any activities that you need to do safely. This means that for each activity you will carry out, you need to use all the mitigation actions shown on the risk assessment. These may include, for example, wearing appropriate personal protective equipment, following safe working procedures and using machine guards.

Do not use any equipment that is damaged or incomplete. For example, don't use electrical equipment such as a soldering iron if the wires to it are frayed or damaged; don't use a machine if its guard is cracked. Tell whoever is responsible for supervising the practical work of the problem immediately.

Always clean away any mess that you make. This includes returning any tools and materials to the correct places.

Objectives

- Explain what features indicate high-level skills.
- Explain the need to use appropriate tools and methods.
- Explain why quality control is needed to achieve high-quality products.

Examples of making projects

Danielle has made a target game for use at a school fair, Figure **A**. It is a very good example of a product with high-level making skills. The PCB has been produced from a CAD drawing by photo etching. The acrylic case has been produced from a CAD drawing using a CAM laser cutter to give a highly accurate finish. The inputs and outputs are precisely assembled and securely held to the case. There is access to the battery compartment for the battery clip wires. Strain holes have been used to secure wires. Figure **B** shows the underside of the PCB. It has thick tracks and pads, which aids the soldering process and shows high-level skills. The PCB also indicates the positioning of many of the components that populate the board.

Danielle's electronic product looks as though it was produced with the appropriate tools, materials and technologies in a safe and skilful way. It is a rigorous and demanding outcome which is suitable for its target market.

Jon's infrared controlled buggy has excellent electronics and PIC programming, which allows it to work perfectly and consistently. With the time allowance for making and the deadlines for completion of controlled assessment, it was a sensible decision to model the buggy from Lego® and buy a manufactured case to house the IR control unit. As shown in Figure **C**, it still meets its target of negotiating a track 'remotely' – a lot of fun for young children at the school fair. It is still expected that batteries, PCBs, switches and other inputs and outputs should be securely located.

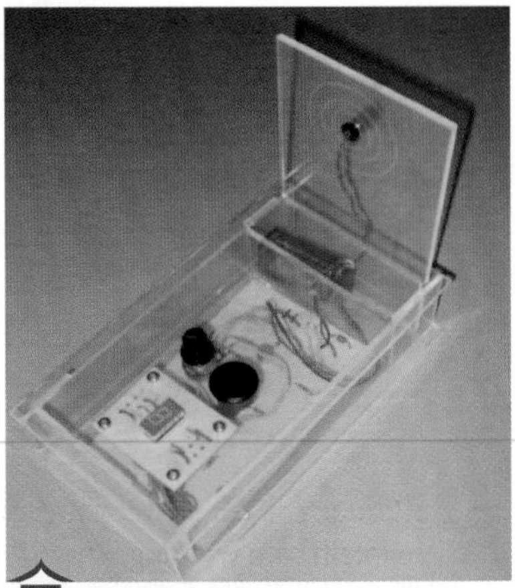

A *Danielle's target game*

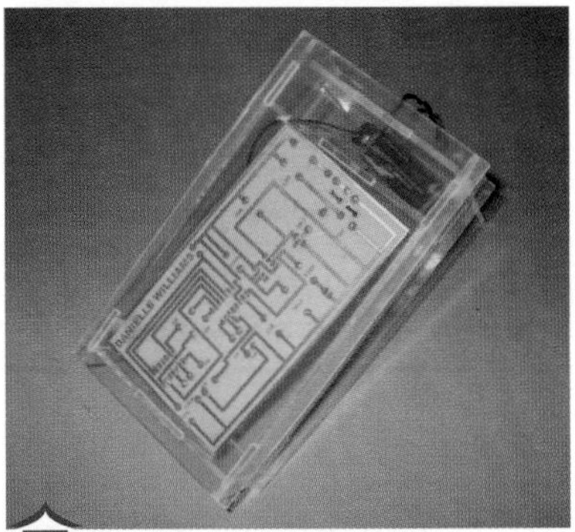

B *Danielle's target game PCB*

Rishi's cat flap project has a lot to recommend it in terms of its quality of manufacture, Figure D. The internal layout is well organised, with the PCB, battery and speakers held in position securely. Heatshrink has been used to insulate the wires attached to the microswitch. The vacuum-formed case is colourful, with fun and appropriate graphics. There is attention to detail in the drilling of the speaker holes. The jack and socket for the cat flap micro switches tidy up what could otherwise be a tangle of wires.

AQA Examiner's tip

Remember that you are being observed during practical work. The way in which you work will affect your grade and future employment.

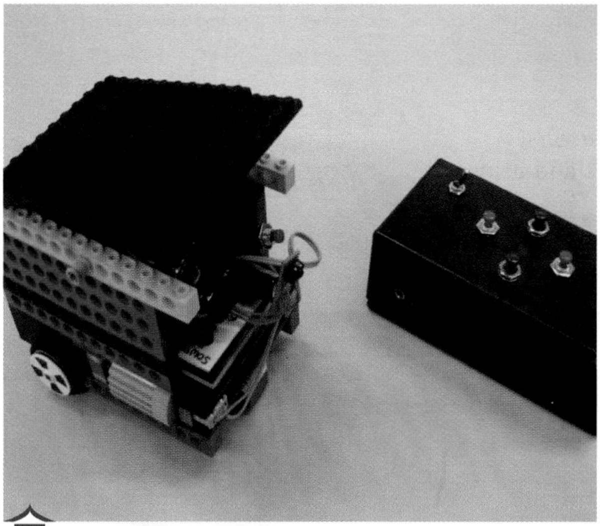

C *Jon's buggy*

D *Rishi's cat flap project*

Activities

1. Using the internet, find out about the Health and Safety at Work Act. Prepare a presentation summarising:
   - the responsibilities of the employer
   - the responsibilities of the employee
   - the role of the Health and Safety Executive
   - who can be prosecuted under the Act, and what the maximum penalties are.
2. Produce an A3 poster that could be used in a school workshop to explain safe working.

Remember

Some common causes of accidents in the workshop

Carelessness or loss of concentration

Not following safety rules

Lack of training in using the equipment

Unguarded or badly maintained equipment

Tiredness and fatigue

Slippery floors and messy work areas

Poor behaviour – not working sensibly

Trying to lift items that are too heavy

Summary

High-level making skills include:

an enclosure that is well made and matches the product's function

inputs, outputs, PCBs and batteries neatly and securely located

a well-designed PCB which fits the enclosure, and features such as strain holes for battery clips

good soldering, no loose wires, sleeving of wires

quality control – attention to detail, accuracy, colour coding of wiring.

kerboodle!

8.14 Evaluation of the finished product

Functional testing

The first step in evaluating the product is to check that it works as intended. This is called **functional testing**. This normally means trying it out. This could also mean running tests – for example, in industry some manufacturers of chairs have a machine which simulates a large person sitting on the chair 10 000 times or more to check that it does not fail over time.

Comparing the product to the specification

Whilst functional testing checks that the product works as intended, it may not cover all of the needs that the product must satisfy, such as cost, aesthetic appeal or recyclability. To ensure that all the relevant needs are evaluated it is necessary to test every need in the specification. This will probably mean that a number of different tests need to be carried out. Where possible, the testing should be objective. This means that it should be based on facts and numbers, rather than subjective opinions. Table **B** shows an evaluation against the specification for the keypad controlled alarm.

Using the evaluation to improve the product

During the evaluation you might identify some needs in the specification which have not been satisfied. In this case, you must explain why these needs have not been met and what could be improved to allow them to be satisfied.

You can also use the evaluation to identify how well certain needs were met. If there is the potential to make improvements that would allow the product to meet the needs better, these should also be highlighted. One way to do this is to include some pictures of your finished product, labelled to show key features, any modifications made during the design and making process, and possible improvements.

Objectives

Explain why it is important to compare the product to the specification.

Explain how the evaluation could be used to further improve the product.

Key terms

Functional testing: testing to check that the product does what it is meant to do, sometimes carried out by actual use.

A *A finished project*

B *Example of an evaluation against the specification for a keypad controlled alarm*

No.	Need	How tested	Result of testing	Outcome
1	The product is a keypad controlled alarm system. It requires a 3-digit code number to be entered to turn off the alarm.	Testing that the key code turns off the alarm (and that wrong key codes do not)	Correct code deactivated alarm. Three wrong codes didn't	Pass
2	The product should be controlled by a picaxe-20M microcontroller.	Checking the components used and reprogramming the chip	Pass	Pass
3	The product should be linked to two contact sensors and two motion sensors, using multi-core wire.	Checking the components used and checking that each sensor worked, by activating the alarm	Each sensor activated the alarm	Pass
4	It should include five LEDs, to show whether the alarm is turned on and which sensor has been activated.	Checking that each of the LEDs works by activating each sensor in turn	Each LED activated by the correct sensor	Pass

5	The control box for the product should be powered by three 1.5 V AA batteries and include the keypad.	Testing that the keypad deactivated the alarm	Correct code deactivated alarm	Pass
6	It should include a relay to operate a siren, which can be heard up to 30 metres away.	Testing that the siren can be heard while standing 30 metres away	Heard from 30 metres away	Pass
7	The siren should be housed in a separate enclosure and powered by four D batteries.	Visual check	Pass	Pass
8	It should include an on-off switch and a download socket, so that it can be reprogrammed.	Reprogramming the system and turning it on and off	Pass	Pass
9	The enclosure for the control box should be made of white HIPS, with a maximum size of 150 × 250 × 100 mm.	Measuring the enclosure with an engineering rule	140 × 95 × 25 mm	Pass
10	The enclosure for the siren should be made of yellow HIPS, with a maximum size of 150 × 250 × 100 mm.	Measuring the enclosure with an engineering rule	140 × 200 × 94 mm	Pass
11	The PCB and batteries for the control box must be secured within the enclosure so that they cannot move around. The siren and batteries must be secured within their enclosure so that they cannot move around.	Visual check that the parts were held in place. Then shook it for 2 minutes listening for any rattling or loose parts and checked again	Parts held in place	Pass
12	It should be possible to open both enclosures to change the batteries. However, it must be possible to close them to prevent water entering.	Dipping the closed cases in water for 5 minutes and seeing whether any water gets in	No water inside	Pass
13	There must be no sharp edges on the enclosures.	Carrying out a silk test on the model of the case	No snags	Pass
14	The product must be suitable for one-off production.	Making models and prototyping the circuit on a breadboard	Pass	Pass
15	The total cost of the parts used must be less than £10.00.	Adding up the cost of the parts used on the materials list	£8.13	Pass

Activity

Create a list of the tests that might be carried out during the testing of an electric kettle. Using these tests, work out what the specification for the product was.

Summary

Functional testing should be carried out to make sure that the product works as intended.

The product should be compared to the specification to check that it meets all of the identified needs, including those that do not relate directly to the function.

If there are any needs in the specification that are not satisfied, these must be explained. You should also identify how the product could be further improved.

AQA Examiner's tip

It is essential to compare the finished product to the product specification.

Another way of demonstrating how you have modified or improved your design is to keep a practical diary, with a photograph of each stage of the manufacture and a section that describes and explains any changes or modifications made to the design.

kerboodle!

???? Your questions answered

How long should the project be?

The project should take approximately 45 hours in total. Half the time should be spent on design work, the other half on the practical. The design folder should comprise of a total of approximately 30 sides of A4, or the A3 equivalent.

Can I do any of the work at home?

Your project will be closely supervised by your teacher, with most of the work being carried out in school. This will mean that there will be restrictions upon what work you are allowed to complete at home and your homework tasks are likely to be related more to planning what you will need to be effective in the next D&T lesson or related to the written paper. It will, of course, mean that for every lesson you will need to ensure that you are making full use of the time available.

Can I write my own design brief?

No. You must use a brief which has been set by the Awarding Body (AQA) and given to you by your teacher. You may be given several briefs to choose from.

What does target market mean?

The target market is a specific group of people for whom your product is designed. It could be toys for children, fancy dress for partygoers, but whoever it is you must take their thoughts and needs into account.

How much research should I do?

You should not need to do too much research, perhaps as little as two to three sides. You should ensure that it is relevant to electronic products. If you need to do more, it might be best to summarise your research and acknowledge the source.

Is analysis necessary?

Analysis is very important! Properly done, it allows you to identify areas of research and leads to the generation of effective and innovative ideas. It is also a key to writing the initial specification.

How do I record my design ideas and how many should I have?

Ideas for the electronics can be recorded by sketching or computer generating circuit diagrams, as can designs for the enclosure. Concise notes can be added for further explanation. There is no exact number of ideas that need to be generated, but you will need enough to be able to make a reasoned choice for your solution – three ideas each for circuit and enclosure would be sufficient.

How do I develop a product?

After you have chosen your solution, development takes place by:

- investigating and experimenting with electronic components to make your circuit as efficient as possible
- creating and refining a program for use with PIC's
- experimenting with the shape of the PCB to make it match the product's enclosure
- investigating whether the design for the product is sustainable and how to improve its sustainability.

Does my product have to work?

Electronics requires a high level of precision. Teachers know that the smallest of inaccuracies during the development and manufacture of a product can stop it working, so the principle is that if a product has been designed correctly and it has worked at the modelling stage, a candidate should not be penalised if it does not work as a finished product. However, this does not mean that poor quality of manufacture can be ignored.

How do I achieve high marks?

The product should be fit for purpose and reliable. For electronic products, this means a PCB with wide tracks and big tabs, good soldering, sleeving of components, correct use of single- and multi-core wire, battery and PCB located securely. The enclosure should be well made, showing accuracy and attention to detail with a neat and tidy appearance. Quality assurance and control should be evident in the design and making, with attention given to health and safety and risk assessment.

Who will mark my controlled assessment?

Your teacher will mark your work when it is complete. The marking must be carried out before the exam board deadline. These marks will be sent to the exam board and its moderator. The moderator will request some or all of the candidates' work and moderate it to try and agree with the school's assessment. When this process has been completed results of the the moderation is sent to the exam board.

Glossary

A

ADC: analogue-to-digital converter.

Alloy: a mixture of two or more metals.

Amp: unit of measurement of current.

Amplifier: a device that can increase the output in proportion to the input.

Analogue: of a signal that is variable (that is, does not just have two states of on or off).

Analysis: reviewing the research and deciding what it means for your product.

AND gate: a logic gate that requires both inputs to be high to produce an output.

Anode: the positive leg of an LED.

Astable: a circuit that provides a pulsing output signal.

Automate: use computer-controlled machines instead of workers to perform tasks.

B

Back emf: a momentary reverse flow of electricity when an electromechanical component is switched off.

Batch production: making a quantity of parts before switching to making another product.

Battery: a component or unit which stores electrical energy chemically.

Bending: forming an angle or curve in a single piece of material.

Bias: the voltage (0.6 V) required to allow a transistor to be switched.

Binary: number system used in digital devices with only two possible values for each digit, 0 and 1.

Bistable: a circuit which stays on after a momentary signal is received to the input. Also known as a latch or a flip-flop.

Breadboard: a commonly used name for a prototype board.

Breadboarding: a method that allows you to produce temporary circuits that do not require components to be soldered.

C

Capacitor: a component that stores charge.

Cathode: the negative leg of an LED.

Ceramic: an inorganic material, normally an oxide, nitride or carbide of a metal.

Circuit: an assembly of electrical or electronic components that exists to perform a function.

Client: the person that the work is being carried out for.

Clipping: distortion to the signal caused when it is amplified beyond the voltage of the power supply.

Closed loop: a system that can alter its output based on feedback.

CMOS: complimentary metal oxide semiconductor.

Communicate ideas: share a concept with others.

Comparator: a device that compares two values.

Component: an individual part

Composite: a material that is made from two or more material types that are not chemically joined.

Computer numerical control (CNC): using numerical data to control a machine.

Computer-aided design (CAD): the use of computer software to assist the design of a product.

Computer-aided manufacture (CAD): using computers to operate machines to produce a product.

computer integrated manufacture (CIM): the use of CAD to design a product which is then transferred directly to be made on a CAM machine.

Conductor: a material that electricity can pass through, such as the wire or track used to connect components.

Constraints: things that limit what you can make.

Conventions: rules of presentation that drawings must conform to.

Critical path: the shortest route through the timing plan, where each step contributes directly to the lead time.

D

Darlington pair: a combination of two transistors, normally in the same case.

Decade counter: a counter that outputs the results in base 10.

Decoder: a device that converts binary into decimal format.

Design brief: a short statement of what is required.

Design parameters: the values for characteristics that the design has to satisfy.

Digital: of a signal that has only two states: high (on) or low (off).

Dimensions: sizes.

Diode: a component that allows current to flow in one direction only.

Duration: the length of time from the start of one pulse to the start of the next.

E

Electrical: electrical components are simple conductors and perform a function when electricity flows through them.

Electrolytic capacitor: a capacitor that is polarised and will only work if attached the correct way round.

Electronic: electronic components are devices that include semiconductor materials in an electrical circuit.

Enclosure: a container for an electronic device. This includes cases, graphic displays, garments and soft containers.

F

Features: details of the design.

Feedback: information from sensors used to modify the output of a system.

Ferrous metal: a metal that contains iron.

FET: a field effect transistor.

Fibreglass: a non-conductive composite material made from glass fibres and plastic resin.

Flowchart: a diagram showing a sequence of operations or activities.

Form: the size and shape of the product.

Frequency: the number of pulses per second, measured in Hertz.

Function: what the product is intended to do.

Functional testing: testing to check that the product does what it is meant to do, sometimes carried out by actual use.

G

Gain: the amount of amplification provided by a transistor.

Gantt chart: a type of timing plan.

Grain: the direction or pattern of fibres found in wood.

H

Hardwood: a wood from a deciduous tree.

Hazards: things that cause a risk of harm or injury.

Heat sink: a metal plate used to dissipate heat.

High: a digital state also known as 1 or on.

High-volume production: making large numbers of parts using dedicated machines.

Hysteresis: the time lag between a correction being made to a system and the output of the system returning to the target value.

I

IC: integrated circuit.

Impedance: resistance.

Injection moulding: the process of making plastic parts by forcing liquid plastic into a mould and allowing it to solidify.

Insulator: a material that does not allow heat or electricity to pass freely through it.

Inverter: a digital device that turns a high input into a low output and vice versa.

L

Laser cutting: using a laser to cut out a shape by melting or vaporising the material along the cut line.

Latch: a device that maintains its switched position.

Layout: the design of the tracks on a PCB.

LDR: light-dependent resistor.

Lead time: the amount of time needed to complete an activity or to supply a product.

LED: light-emitting diode.

Logic probe: a device that can be used to determine whether a digital signal is high or low.

Low: a digital state also known as 0 or off.

Lower threshold: the voltage level below which a digital component recognises an input as low.

M

Maintenance: carrying out activities to extend the usable life of a product.

Manufactured board: a wood product made by processing or pulping wood particles or sheets.

Mark time: the time that a pulse output is high.

Mark/space ratio: the balance between the time a pulse is high and the time it is low.

Marking out: drawing the design of a part onto the materials that it will be made from.

Mask: a pattern used to shield areas of the photosensitive PCB from light.

Microcontroller: a type of programmable microprocessor.

Mitigation actions: precautions taken to reduce a hazard.

Modelling: simulating the use of a circuit or product.

Momentary: switches off after being operated.

Monostable: a circuit that produces a single output for a fixed period of time.

Mould: a former used to shape a part.

Multimeter: a device that can be used to measure current, voltage or resistance.

N

NC: normally closed.

Negative feedback: returning part of the output signal to the inverting input.

NO: normally open.

Non-ferrous metal: a metal that does not contain iron.

NOT gate: a logic gate that inverts the input to produce an output.

O

Ohm: unit of measurement of resistance.

Ohmmeter: a device used to measure resistance.

One-off production: making a single product or prototype.

Op amp: operational amplifier.

Open loop: a system that is set to achieve a required value.

Opto-isolator: a light-based interface device.

OR gate: a logic gate that requires only one input to be high in order to produce an output.

Orthographic drawing: a working drawing of a part showing three views, to communicate the dimensions of the design.

Oscilloscope: a piece of equipment with a visual display that can be used to accurately measure rapidly changing signals.

P

Pad: a contact point for a component.

Photochromic: changing colour with changes in the level of light.

PIC: peripheral interface controller; or programmable interface controller, programmable integrated circuit.

Piezoelectric material: a material which changes shape fractionally when a voltage is applied to it.

Pinout: a diagram that shows what each pin does.

Planned obsolescence: designing a product for a limited life.

Polarised: having a positive side or leg and a negative side or leg.

Pollution: contamination of the environment.

Polymer: an organic material made up of a chain of single units called monomers.

Potential divider: a device that divides a voltage so that its output voltage is some proportion of the input voltage.

Preferred values: commonly available resistor values.

Printed circuit board (PCB): the specially-designed base that an electronic circuit is assembled on using soldering.

Production plan: the instructions on how to manufacture a product.

Program: the series of instructions that tell a microcontroller what to do.

PTB: push-to-break.

PTM: push-to-make.

Pull down resistor: a resistor used to tie down inputs to prevent false readings due to static electricity.

Pull up resistor: a resistor used to ensure an input receives the full voltage of a supply.

Q

Quality assurance (QA): taking steps before making a product to make sure that it is made correctly.

Quality control (QC): checking that a part or product is correct after it has been made.

Quantifiable: measurable.

R

Rapid prototyping: making an example of a product for evaluation, normally on a single computer-controlled machine.

Real-world: actual experience or practice, as against a virtual or theoretical object.

Recycled: made using materials that have been used before and reprocessed.

Reference voltage: a voltage value set by the user.

Relay: an electromagnetic interface device.

Rendering: applying colour or texture to a sketch or drawing.

Research: activities investigating or clarifying the design needs.

Resistor: a component that limits current.

Risk assessment: a review of the potential of an activity to cause harm.

S

Scale: the ratio of the size of the drawing to the size of the part.

Sensitivity: the amount a sensor's output changes with changes in the phenomenon being measured.

Sequence: the order in which a series of steps need to be carried out.

Series: an orientation where components are located end to end.

Seven-segment display: visual display arranged to show decimal numbers.

Shading: creating different tones on a sketch or drawing.

Shape memory alloy: a metal that, once deformed, will return to its original shape when heated above its transition temperature.

Sinking: current flows into pin 3 of the 555 IC.

Sketching: a quickly produced visual representation of an idea.

Softwood: a wood from a coniferous tree.

Sourcing: current flows out of pin 3 of the 555 IC.

Space time: the time that a pulse output is low.

SPDT: single-pole double-throw.

Specification: a list of needs that the product must satisfy.

SPST: single-pole single-throw.

Strip board: a polymer or composite board coated with strips of copper, with holes at predrilled intervals.

Surface mount: an approach where components are positioned on the surface of the PCB.

Switch bounce: when mechanical switches make multiple (unintended) contacts whilst being used.

Symbol: a drawing used to represent a component.

System: a collection of parts that interact with their environment and perform a function.

Systems diagram: a schematic representation of a system.

T

Thermistor: a resistor where the resistance changes with temperature.

Thermochromic: changing colour with temperature.

Through hole: an approach where components need to be put through holes in the PCB.

Thyristor: a component often used as a bistable or latching device.

Tie down: link to the 0 V supply.

Time constant: the time to charge a capacitor used in series with a resistor.

Tolerance: the possible variation in the accuracy of a resistor's or capacitor's value.

Track: a path of copper, joining components.

Transducer driver: a device such as a transistor or thyristor that can provide a high power output.

Transistor: a component that functions as an electronic switch and amplifier.

Trigger voltage: the voltage needed to turn on a thyristor.

Truth table: a chart that explains the relationship between the inputs and the output of a logic gate.

U

Upper threshold: the voltage level above which a digital component recognises an input as high.

User needs: the things that the customers require the product to do.

V

Vacuum forming: forming a thermoplastic sheet over a mould, using heat and a vacuum.

Virtual modelling: a computer simulation of a system that enables a user to perform operations on the simulated system.

Voltage signal: output voltage from a potential divider.

Index

WALES, NEWPORT LIBRARY AND INFORMATION SERVICES CAERLEON